The Emergence of Somatic
Bodymind Therapy

Critical Theory and Practice in Psychology and the Human Sciences

Titles include:

Barnaby B. Barratt
THE EMERGENCE OF SOMATIC PSYCHOLOGY AND BODYMIND THERAPY

Derek Hook
FOUCAULT, PSYCHOLOGY AND THE ANALYTICS OF POWER

Mary Watkins and Helene Shulman
TOWARD PSYCHOLOGIES OF LIBERATION

Critical Theory and Practice in Psychology and the Human Sciences
Series Standing Order ISBN 978-0-230-52113-1 (hardback) and 978-0-230-52114-8
(paperback)
(*outside North America only*)

You can receive future titles in this series as they are published by placing a standing order. Please contact your bookseller or, in case of difficulty, write to us at the address below with your name and address, the title of the series and the ISBN quoted above.

Customer Services Department, Macmillan Distribution Ltd, Houndmills, Basingstoke, Hampshire RG21 6XS, England

The Emergence of Somatic Psychology and Bodymind Therapy

Barnaby B. Barratt
Senior Research Fellow, University of Cape Town

First published 2010 by
PALGRAVE MACMILLAN
Paperback published in 2013

Palgrave Macmillan in the UK is an imprint of Macmillan Publishers Limited, registered in England, company number 785998, of Houndmills, Basingstoke, Hampshire RG21 6XS.

Palgrave Macmillan in the US is a division of St Martin's Press LLC, 175 Fifth Avenue, New York, NY 10010.

Palgrave Macmillan is the global academic imprint of the above companies and has companies and representatives throughout the world.

Palgrave® and Macmillan® are registered trademarks in the United States, the United Kingdom, Europe and other countries

ISBN: 978-0-230-22216-8   hardback
ISBN: 978-1-137-31096-5   paperback

This book is printed on paper suitable for recycling and made from fully managed and sustained forest sources. Logging, pulping and manufacturing processes are expected to conform to the environmental regulations of the country of origin.

A catalogue record for this book is available from the British Library.

A catalog record for this book is available from the Library of Congress.

*May all beings be joyful and free;
may these writings contribute
to the joyfulness and freedom of all beings.*

This page intentionally left blank

Contents

Acknowledgements

It is impossible to acknowledge properly all those who have, directly or indirectly, contributed to my writing of this book, for the list would include not only my academic and psychoanalytic teachers, but also those who have fostered my somatic and spiritual growth. I am in gratitude. More specifically, various friends and colleagues read and commented on portions of the manuscript. These include: Susan Aposhyan, Christine Caldwell, Randy Earnest, Axel Hoffer, Don Hanlon Johnson, Sam Kimball, Peter McLaren, Jerry Piven, Tod Sloan, Mary Watkins, and Lloyd Williams. I thank them (and, as is customary, I immediately add that any errors and all peculiarities in this text are no one's responsibility other than my own). Robert Romanyshyn was kind enough to suggest some Jungian references; Judyth Weaver also pointed to some important somatic literature; John Leavey guided me to some crucial writings on the body in relation to deconstructive method; and Kay Campbell was kind enough to suggest some perspectives on the significance of touch in infancy. Denise Dorricott and Marsha Rand skillfully edited various portions of the manuscript. And to Marsha, I owe a special debt of gratitude for all her support.

Section I
Introducing a New Discipline

Our world is changing rapidly, and our understanding of what it means to be human and of the place of humans in the universe is shifting. Ways of thinking that have governed the Western world for the past several hundred years are now being radically subverted, proving themselves to be limiting or inadequate. These changes hint at new ways by which we might understand ourselves and our world. But often we do not yet quite know what the new ways of thinking will be or how they will affect our sense of the human condition and the planet we inhabit. Often, we just know that the thinking that has preceded us is not sufficient. Understanding the human condition — some would say understanding the human *psyche* or "soul" — is the mandate of psychology and perhaps of all the social and biological sciences associated with it. This book is about new ways of thinking in psychology.

In this first section, and throughout the book, we will consider the nature of change in the history of ideas that constitute the discipline of psychology. We will also locate this investigation within a more general consideration of the nature of change in our understanding of the processes of knowing and being within the development and the diversity of human cultures. This will include a discussion of changes in what the modern western world considers *scientific* and what it condemns as unscientific.

Here you are invited to assess the significance of a newly forming discipline within the human sciences, and along with it a group of healing practices that are both newly emerging and re-emergent; practices that are derived from this new discipline, yet are also derived from ancient traditions of wisdom that are currently being remembered or rediscovered. The discipline is *somatic psychology* and we will name these diverse healing practices *bodymind therapy* (although the group has often been called "body psychotherapy," and sometimes body-mind psychotherapy or body-centered psychotherapy). In this context, bodymind therapy is the applied aspect of somatic psychology. As this book unfolds, I hope that it will prompt you to reopen your vision of the psyche in relation to the human experience of embodiment; that you will realize the significance of the contemporary emergence of somatic psychology and bodymind therapy as an indicator of the profound change that is occurring in our most fundamental understanding of what it means to be human. So as we proceed together, we will initially discuss the nature of change, and then gradually focus on the specific topic of interest to us.

This first section of the book, with its five chapters, will set the stage for this assessment of the significance of this disciplinary venture — the emergence of somatic psychology and bodymind therapy. The second section will offer you an account of what I consider the seven main sources that contribute to the contemporary budding and blossoming of this discipline, and the final section will present several discussions of current challenges in this field. The impetus of this book is to empower you to consider questions such as the following:

- Is somatic psychology generating excitement simply because it is a newly formed sub-discipline within the general field of psychology (a field which developed so expansively through the twentieth century)?
- Do the practices of bodymind therapy merely comprise a powerful new branch or novel application of the familiar field of psychotherapy (and the technology of "behavior change"), which unfolded so dramatically through the twentieth century?
- *Alternatively*, is it possible that somatic psychology and bodymind therapy are the harbinger of a radically different future? Do they perhaps intimate a profoundly different way of understanding and appreciating the human condition, constituting an emergent and revolutionary break with the psychology and the psychotherapeutic methods that dominated the twentieth century?

With any comparatively recent cultural or scientific venture, it is difficult to know what it means to call something "new." Predictions of "revolutionary significance" usually need to be treated with healthy skepticism. After all, contemporary culture, impelled by entrepreneurial capitalism, is extraordinarily faddish and prone to transient fashion. Corporations make spurious claims about how some "new" product will change our lives and clearly they promote illusions of novelty merely to stimulate the market demand for whatever they are selling. When a "revolution in hair styling" is declared, or a "new line of dietary aids" is depicted as indispensable, or the latest line of resource-consuming gadgetry is declared "the solution of the future, here today," there is every reason to cease paying attention. We are advised to ignore such claims because there is every reason to believe that the "same old stuff" is merely being peddled as if it were genuinely a means by which we might break out of the deleterious repetitiveness of our lives. Yet despite our culture's faddish promotion of the illusions and the fetishes of change, there is a profoundly different sense in which incessant change is a reality. Shifts and changes do occur genuinely in our culture and in our sciences, and it is especially challenging — but perhaps not entirely impossible — to assess and appreciate the significance of a cultural or scientific trans-formation as it is in the process of emerging.

When rock-and-roll burst upon the scene between the late 1940s and early 1960s, it could justifiably be hailed as a "new" musical genre and indeed it has come to have a massive and unprecedented social impact on cultures across the globe. This new genre, defined by its use of the electric guitar as well as its particular rhythms and accentuated back-beat, did not emerge *de novo*. Rather, it had its sources in jazz, blues, gospel and diverse forms of folk and country music, including boogie woogie, jump blues, and western swing. It became its own distinctive genre (and allegedly was given its own name in 1951), and it burgeoned through the now classic period of the 1950s and 1960s. The point here is that, throughout this dramatic period of its emergence, no one could have surely predicted the extent or character of its eventual worldwide impact — an impact that has revolutionized not only the realm of music, but that has colorfully influenced almost every aspect of human culture.

When something "new" surfaces in the scientific domain, it is some-times a new discovery within an established field of investigation, and sometimes it is the opening of an entirely new field. When the pro-cedures to produce synthetic polymers were developed in the late 1930s, the utility of nylon fibers could quickly be predicted (as a sub-stitute for silk and for many other uses). Indeed, the advent of thermo-plastics has had an inestimable impact on the functioning of our lives. But such discoveries did not change the principles of chemistry as a science, nor did they redefine the underlying philosophy of the field.

A somewhat different sort of novelty would be the emergence of an important and previously unavailable application of familiar scientific principles to an entirely new topic or area of investigation — the inauguration of a new discipline or field of inquiry. The case of molecular biology provides an example. Although it had antecedents in eighteenth century microscopy, the discipline was only inaugurated in the 1930s. Its specific mission to study the vital role of nucleic acids and proteins only became possible when X-ray diffraction, electron microscopy, and other technologies became available. With this availability, a new discipline with its own principles and methods was established.

Beyond these specific events, the history of science shows us that sometimes the emergence of apparent novelty is not just a particular line of discovery or the opening of a previously unavailable field of investigation. Sometimes, when something "new" emerges in science, it comprises such a profound shift in our way of appreciating ourselves and the universe that we think of it as a revolutionary change in our way of thinking about what it means to know and what it means to be wise.

Such was the case with the Copernican revolution. Although Vedic, Hellenic, and Islamic philosophers had conjectured the possibility of a heliocentric universe, Nicolaus Copernicus' demonstration of the movements of celestial objects at the beginning of the sixteenth century, along with Galileo Galilei's support for Copernicanism at the beginning of the seventeenth century, entirely changed not only western perspectives on earth's place in the cosmos, but contributed to a radically revised understanding of the significance of being human. By the time of Isaac Newton's publication of his *Philosophiae Naturalis Principia Mathematica* in 1687, the Copernican revolution had also radically altered our understanding of the nature of science itself. In this sense, an entirely new way of comprehending the universe opened up for us. Our ways of knowing and being were revolutionized in that our fundamental assumptions about the nature of humanity and divinity shifted. This was a "novelty" that took about two hundred years to emerge fully, but it profoundly changed the way we think, and rocked every aspect of the world in which we live. Western culture's relinquishment of the patterns of medieval thinking, and the emergence of the modern era, constituted what Michel Foucault would call an "epistemic shift."

It is in this context that this book invites you to entertain the question: *Just how radical are the implications of the emerging discipline of somatic psychology and the accompanying healing practices of bodymind therapy?*

There is a range of possibilities. At one end, somatic psychology might be understood simply as a broadening of the scope of psychological science to include a new topic or area. After all, the science of psychology itself expanded dramatically in the west at the end of the nineteenth century, but it was not until the 1960s and 1970s that the principles and research methods of psychology were systematically applied to the behavior of people in athletic activities, and the sub-discipline or specialization of "sports psychology" was inaugurated (although there had been some earlier work by pioneers such as Norman Triplett, Carl Diem, and Coleman Griffith). If the emergence of somatic psychology is comparable to this inauguration, then perhaps the new discipline is merely an interesting application of familiar principles and research methods to a new topic — the "new" topic being the human experience of embodiment.

Similarly, the clinical practices of bodymind therapy or "body psycho-therapy" might be understood simply as the addition of a set of new techniques to the familiar practices and procedures employed by psycho-therapists. After all, when family therapy (or "family systems therapy") started to become popular in the mid-twentieth century, it constituted not only an extension of psychotherapeutic practice, it also began to offer a wide variety of new techniques and clinical perspectives by which to address psychological suffering (with pioneers such as Nathan Ackerman, Carl Whitaker, Murray Bowen, Jay Haley, and Salvador Minuchin). If the popularization of bodymind therapy were to be com-parable to that of "family systems," then perhaps it is to be appreciated merely as an exciting application of new clinical methods that can be appended to the familiar goals and general methods of clinical psychology.

At the other end of the range of possibilities, the significance of somatic psychology and bodymind therapy may be far more radical than this. *Perhaps the emergence of somatic psychology and bodymind therapy portends a profoundly different and potentially revolutionary way of appreciating the human condition.* In this book, you are invited to enter-tain this possibility. You are invited to consider whether somatic psychology might indeed imply a radical critique of the "psychology" and "psychotherapy" that dominated the twentieth century, and whether indeed it might prove to be the wave of the future.

In this first section Chapter 1 will define somatic psychology and bodymind therapy in relation to the modes of psychology and psycho-therapy that dominated the twentieth century. Chapter 2 will suggest the significance of the emergence of this new discipline within the epistemic shifting that seems to be occurring in contemporary science and culture — the implosion of the modern era and the intimations of a postmodern world. Chapter 3 will describe just a few of the practices of bodymind therapy in order to give you the flavor of its healing potential. Chapter 4 will amplify these descriptions by discussing some of the principles of healing that underpin bodymind therapies (as well as the ideological controversies that surround the issue of genuine healing *versus* social adaptation). And Chapter 5 will survey the current disciplinary state of somatic psychology and bodymind therapy as far as they have developed since the closing decades of the twentieth century.

Psychology at the Crossroads

Somatic psychology? Isn't that an oxymoron? A contradiction in terms? After all, 'soma' refers to the body, the material aspect of being human, and 'psyche' refers to the soul or the mind, the non-material aspect of our thoughts, feelings, willpower, or spirit, which seems to be housed in the brain but that isn't wholly reducible to the stuff of neurons ... soma and psyche are two different sorts of being. Every undergraduate since Descartes knows that!

Indeed, every student in the western tradition "knows" that there are at least two aspects to the "beingness" of being human — the *body* and *mind*. And every student knows that often a third part, the ineffa-*soul* which somehow lives beyond the individual's lifespan, is axiologically differentiated from these two (as a topic of conjecture, faith, or metaphysical experience). René Descartes' version of this dualism, which had antecedents in pre-Aristotelian and Avicennian philosophies, merely formalized the prevailing wisdom of the early seventeenth century. Its influence persisted through the twentieth century and continues today (although the burgeoning dispute over its tenets makes this persistence increasingly shaky). The body, *res extensa*, operates as a machine — having the material properties of spatial extension and motion that obey the laws of physics. By contrast, the mind, *res cogitans*, has neither extension nor motion and is not ruled by physical laws. In this cosmology, there is a mind-body connection, but its operation remains enigmatic. Thoughts are structured by time, but do not occupy space. Thus, there is an absolute, but problematic, divide between the immaterial mind (housed somehow in the cortex) and the material body. The mind controls the body — at least as best it can — but how it does so remains quite obscure, even to the most dogmatic Cartesian. Sickness may compromise the effectiveness of this governance, and death, of course, terminates the relationship. In health, the body can still influence the mind, notably during those ruptures of rational masculinist thinking that were derogatorily called "passion." All this characterizes the vision of the human condition that Descartes bequeathed us.

Subsequent to Descartes' writings, the nature of the mind-body connection has been a subject for heated debate from at least the mid-seventeenth century to the mid-twentieth century. And even today, the issue of connections between conscious awareness and neuronal functioning provokes vigorous discussion.

However, there is a sense in which, since the mid-1950s, the debates over Cartesian dualism have broken down, or at least radically shifted in the groundwork of their assumptions. In no small measure, this is because technology has offered us the means to examine the concomitancy of experiential awareness and neurological events with a precision that was never previously available. The arrival of such investigatory techniques as electroencephalography, computed tomography, positron emission tomography, magnetic resonance imaging, magneto-encephalography, and transcranial magnetic stimulation have — together with progress in molecular biology and associated disciplines — enormously advanced the sophistication and range of neuroscientific research. But from a broader perspective, it can also be said that Cartesian dualism seems to have exhausted itself because, toward the end of the twentieth century, it became apparent that mind and matter can no longer simply be assumed to be "dichotomously autonomous but somehow connected." We now know that they must be considered in radically different ways from those frameworks of debate that have dominated scientific thinking since Descartes.

These considerations gird the discipline of psychology. So let us now briefly sketch the history of this discipline in order to understand its treatment of bodily matters through the twentieth century.

Pre-twentieth century psychology

Although there are ancient eastern and indigenous traditions of wisdom that can justifiably be described as psychological, *psychology* in the dominant western tradition is perhaps still a relatively youthful discipline. It has its sources in classical philosophy (Greek, Latin, and Egyptian), and later with patristic and medieval scholars such as Aurelius Augustine of Hippo and Thomas Aquinas. Between the eighth and twelfth centuries, there were major contributions from Muslim scholars, such as Isḥāq al-Kindī, Allāh ibn Sīnā (commonly known in the west as Avicenna), Abu Nasr al-Fārābi, and Ibn Rushd (who is frequently called Averroës). Sadly, these contributors are usually neglected in textbook histories of the discipline (Hergenhahn, 2008; Leahey, 2003). Especially between the ninth and thirteenth centuries, there were also many Jewish philosophers who generated what might well be called psychological theory, and were usually influenced by the spiritual practices of Kabbalah. These would include writers such as Abraham Ibn Daud, Solomon Ibn Gabirol, Abraham Ibn Ezra, Ibn Zaddik, Isaac Israeli, and most eminently Moses Maimonides. These contributors have also been seriously neglected, even by some of the more far-reaching and less narrowly conceived texts on the history of psychological thinking (Brock, 2006; Jansz & Drunen, 2003; Lawson, Graham & Baker, 2006).

As a generalization, it can appropriately be concluded that medieval psychology was mostly concerned with issues of the soul. In the meta-theoretical classification promoted by Thomas Teo (2005) and others, it can be categorized as a "metaphysical psychology." In general, it touched on the body only as it pertained to human destiny as defined by our soul. Psychology was, after all, largely under the authority of theological doctrine. This is perhaps less true of Muslim and Jewish scholarship, but it certainly characterizes most Catholic Christian scholars who tended to assume that the body's existence was fundamentally to be understood as an impediment to the soul's ultimate transcendence — its unification with the Christian God (Murphy, 2006).

In the modern era which is usually dated as beginning around the latter part of the sixteenth century, there was a discernible shift from metaphysical to "philosophical psychology" (Teo, 2005). Major influences on the latter discipline would include the work of continental philosophers (starting with Descartes, but including Benedict Spinoza and Gottfried Wilhelm Leibniz), British empiricists and associationists (Thomas Hobbes, John Locke, George Berkeley, David Hume, David Hartley, Thomas Reid, John Stuart Mill, Alexander Bain), early French behaviorists and sensationalists (Julien Offray de la Mettrie, Étienne Bonnot de Condillac), and German philosophical psychologists (Immanuel Kant, Johann Friedrich Herbart, Rudolf Hermann Lotze, Arthur Schopenhauer, and Friedrich Nietzsche). Perhaps with the exception of Nietzsche, the dichotomy of mind and body — under one conceptualization or another — characterizes the thinking of most of these influential writers.

It is not until the late nineteenth century that experimental psychology was launched as a distinctive discipline and, as is well known, it is really not until the early twentieth century that the discipline which is often called — in a hegemonic fashion — *scientific psychology*, expanded dramatically (Plotnik, 2005). Although the proponents of this discipline frequently claimed that only their approaches were truly scientific, this development is more aptly labeled "natural-scientific" (as we shall see). Alongside the growth of this scientific discipline, various clinical endeavors were initiated and the modern mental health movement was expansively established (Porter, 2003; Shorter, 1998). Psychology thus differentiates itself from philosophy, just as psychotherapy differentiates itself from religious preoccupations.

Experimental psychology is usually said to have begun in Europe (and Russia) in the late 1800s with pioneers such as Wilhelm Wundt, Hermann Ebbinghaus, and Ivan Pavlov. William James and Granville Stanley Hall are widely regarded as the major introducers of the discipline in the United States (along with Charles Sanders Peirce). And it is mostly in North America that the inauguration of an experimental approach quickly generated an interest in several kinds of applied psychology. Examples of this are John Dewey's work in education, Hugo Münsterberg's work in industrial relations, and James Cattell's development of Francis Galton's early work on the testing of mental abilities.

Psychology comes into its own in the twentieth century, and it comes to be dominated by two paradigms. These are *cognitive behaviorism* and *psychoanalysis*. As we will see, both have a hostile or neglectful attitude toward the lived experience of our embodiment. However, later in this book I will argue that the pioneer of psychoanalysis, Sigmund Freud, was first and foremost a somatic psychologist.

Around the 1960s, there was then a call for a "third force" in psychology, and the reinvigoration of a "human-scientific" approach to the discipline (Teo, 2005). In some respects, this call surfaced as a protest against the dominance of behaviorist and psychoanalytic approaches to the human condition. In a fundamental sense, certain aspects of this momentum were also indicative of the collapse of old ways of thinking about humanity. In this sense the emergence of a "third force" in psychology is embedded in the turn toward somatic psychology (as will be discussed in later chapters).

At this juncture, we need to review briefly the dominant paradigms of twentieth century psychology to show how they presumed, and inscribed in their theoretical structures, the alienation of mind and body. That is, how they perpetuated Cartesian dualism by assuming this alienation to be the natural state of the human condition.

Cognitive behaviorism

The behaviorist approach to psychology was formulated by John Watson. His 1903 doctoral dissertation from the University of Chicago concerned the "psychical development of the white rat," which became a classic in psychobiology. Later in his career, he worked extensively in the advertising industry, developing campaigns to market cosmetic products throughout the USA. However, his most influential professional contribution was undoubtedly his 1913 essay, "Psychology as the Behaviorist Views It." It is often referred to as the "behaviorist manifesto" (Watson, 1930). This document damned introspective methods, condemned any preoccupation with consciousness, and restricted psychology to the objective study of observable and thus measurable behaviors — an "experimental branch of natural science" (Buckley, 1989; Todd & Morris, 1994). Despite Watson's seminal role, it was the extremist work of Burrhus Frederic Skinner that became better known. Skinner conducted extensive research on operant and respondent conditioning, beginning with rats or pigeons, and then moving on to human subjects. Influenced by Percy Bridgman's operationalist philosophy of science, Skinner spent much of his career as a Harvard professor who argued strenuously that psychology should ignore internal physiology and hypothetical constructs such as the mind (Skinner, 1978). This provoked the influential philosopher, Daniel Dennett (1992, 1997), to parody Marshal McLuhan's 1964 aphorism — "the medium is the message" — and quip that Skinner's message is "there is no medium!" It also inspired the eminent linguist, Noam Chomsky, to write a powerful critique of the entire behaviorist program (Chomsky, 1959). Subsequently, many others have contributed to this line of critique arguing that, although behaviorist psychology is founded on experimentation, its quality as science is dubious considering the ideological restrictions inherent in its definition of the phenomena to be studied.

Skinner's behaviorism focuses exclusively on the manipulation, prediction and control of behavior (Baum, 2005; Mills, 2000). The *experience* of the organism, let alone the subject's bodily experience, is declared irrelevant to psychological science. All that now counts are the external observables — the environmental input to the organism and the behavioral output. "Science" is supposed to be advanced by the ideological exclusion of major segments of the experienced world.

Psychology did not long remain with the dogmatic extremism of Skinnerian externalization. In the early to mid-twentieth century, behavioral psychologists investigating the principles of learning began to debate whether consideration of "internal mediating variables" — cognitive constructs — might indeed be necessary for a fully scientific account of human behavior. Greatly involved in this debate were pioneers such as Edward Thorndike, Clark Hull, Edwin Guthrie, and Edward Tolman. Gradually, the agenda of what is now called "cognitive behaviorism" unfolded, with researchers such as Neal Miller, John Dollard, Hobart Mowrer, Donald Hebb, Julian Rotter, and Albert Bandura being programmatically influential in this development (Rachlin, 1991; Staddon, 2001). By the 1970s, the ideology of cognitive behaviorism virtually annexed academic psychology throughout North America and Europe, to the point where the 1976 President of the *American Psychological Association* could speak with some assurance about the discipline's alleged convergence on the model of cognitive behaviorism and information processing.

Within the realm of clinical enterprises, behavior modification continues to be aggressively advocated as the exclusively "evidence-based" and cost-effective mode of intervention in mental health and other arenas. This is the application of principles of operant and respondent conditioning — the techniques of altering behavioral outcomes by manipulating the contingencies of positive and negative reinforcement. Although Thorndike had anticipated such an application of learning principles, it was not until the 1940s and 1950s that Joseph Wolpe launched behavior modification. This quickly expanded into the industries of "applied behavioral analysis" and "behavior management." These enterprises were popularized in the fields of mental health, education, and organizational psychology, at what critics would consider an alarming rate. As behaviorism gave way to cognitive behaviorism within academic departments of research psychology, Wolpe's style of behavior modification evolved into "cognitive behavioral therapy," with influential figures such as Albert Ellis propounding rational-emotive therapy in the 1950s, and Aaron Beck and Arnold Lazarus developing similar innovations in technique. These innovations have steadily grown in their frequency of clinical usage as well as their sociopolitical influence. Subsequent variants of this basic approach have been called multimodal therapy, cognitive behavioral group therapy, contingency management, dialectical behavior therapy, reality therapy, self-instructional training, solution-focused therapy, and schema-focused therapy (Kring, Davison, Neale & Johnson, 2006).

The so-called "cognitive turn" in psychology and behavioral intervention focuses the discipline on the internal organization — the regulatory structures and functions — of the organism that is necessary to explain its behaviors (Frick, 1994; Fuller, De Mey, Shinn & Woolgar, 1989). In short, the mind is added to the behaviorist agenda. Since the 1970s, the jargon of schemas, scripts, cognitive systems, deep structures, transformational rules, implicit theories, beliefs, personal constructs, decision-making processes, attitudinal sets, and the like, has become the prevailing language of psychological science. It is a language of inner representations of the self and the external world. Although these constructs or structures are indeed *internal* to the subject, they are nonetheless the media of a discipline that continues the ideology of "scientific progress" as benchmarked by the capacity to manipulate, predict and control human behavior. Correspondingly, this ideology upholds a vision of mental health as the behavioral conformity of individuals to prevailing social conditions, and keeps the discipline of psychology very much "in the head."

The focus on cognitive structures — whether theorized as general to the human race, as culturally specific, or as personalized constructs — adheres to Cartesian precepts in the sense that these structures regulate the body and its behaviors, yet are conceptually separate. Perhaps this is slightly less true when, more recently, some cognitive psychology has turned to the study of consciousness. Although here too, "consciousness" is often defined in the limited sense of reflective self-consciousness — the domain of experiences and perceptions that can be articulated in language and that are treated as separate from their embodiment. Programmatically, the stance of the cognitive behaviorist paradigm is that the body is merely the material entity whose behaviors are regulated by external environmental conditions (exclusively in the Skinnerian strategy) and by the agency of internal mental structures.

To the extent that cognitive psychology is ever concerned with the body, it approaches the topic very much in the manner of nineteenth century psychophysics. Its investigations are *about* the body. For example, the lineage of research on body image has produced a sizeable number of investigations concerned with the way in which individuals perceive their own body and the body of others, as well as the attitudes that accompany these perceptions. However, this is precisely not a psychology *of* the body. Rather, it is a psychology *about* the body (a radical distinction that we will discuss in later chapters). To sum this up, although there are some exceptions, it seems clear that the "cognitive turn" in psychological science keeps the discipline ideologically at a distance from the body. Its paradigms effectively reinscribe the alienation of the human subject from the body — the alienation that

Descartes described in his philosophy. The mind, now scientifically described in terms of cognitive structures, is studied in order to explain how it controls the body and its behaviors. In this manner, the lived experience of embodiment is precluded from consideration.

Psychoanalytic psychologies

The paradigmatic program of behaviorism and cognitive behaviorism has thrived on the basis of its experimentalist success in the prediction and control of behavior. These psychologies have thus provided corporations, governments, and the entire military-industrial complex with technologies for human manipulation. These are the agencies that have funded the boom in psychological research as it unfolded throughout the twentieth century.

The ideology of cognitive behaviorism, together with all the research it has produced, has operated powerfully in the service of the dominant sociopolitical order. Against this, it might be claimed that the lineage of psychoanalytic exploration (including the twentieth century tradition of psychodynamics and depth psychology) has upheld a vision of human liberation from the structures of oppression and repression, and thus has established a platform of opposition to cognitive behaviorism. However, this claim is somewhat difficult to sustain because even a cursory examination of the history of psychoanalysis shows how far it has been ideologically compromised (Cushman, 1996; Robinson, 1990; Szasz, 1988). Whatever its radical potential, the alleged progress of psychoanalytic theory and practice throughout the twentieth century is a story in which the discipline, or disciplines, have generally bankrupted their scientific potential and their potential for genuine wisdom in order to promulgate the ideologies of the dominant sociocultural order (almost as much as the forces of cognitive behaviorism).

In subsequent chapters, we will further discuss how the radical potential of psychoanalysis — its capacity to deliver a method of interrogating the psyche in a way that is both scientific and emancipative — has repeatedly been betrayed in favor of theories and practices that perpetuate the ideology of domination. This betrayal is double-edged in that it is a matter of both theory and praxis.

On a theoretical level, it involves abandoning the discovery of the unconscious as dynamically repressed, and this entails abandoning the wisdom of somatic psychology. These factors are intricately connected since whatever is repressed from consciousness is mostly expressed by the "bodily unconscious." This theoretical abandonment of certain fundamental psychoanalytic discoveries about the human condition essentially reasserts the ideology of Cartesian dualism (and we will document this later).

On a practical or methodological level, the betrayal of psychoanalytic inspiration through the course of the twentieth century involves the abandonment of a vision of human liberation in favor of clinical ideologies of cultural adaptation and socially conformist "maturation" (which we will discuss further in Chapter 15). The leading example of this is the pervasive assumption — evident throughout the *Diagnostic and Statistical Manual of Mental Disorders* used by psychiatrists — that "mental health" requires the individual to fit well into the organization of the dominant culture and the ruling social order. Whatever the kernel of radical inspiration inherent in the discovery of psychoanalytic methods, the course of this discipline's development through the twentieth century has all too frequently, and often notoriously, installed its precepts and practices as integral to the ideology of the ruling-class. What has sometimes been called the "Freudian left," the voice of sociopolitical dissent, has always been marginalized within the various strands of psychoanalytic orthodoxy (Robinson, 1990).

Freud was a somatic psychologist — at least from the time he abandoned his neurological ambitions in the very last years of the nineteenth century, until the beginning of the 1914–1918 World War. After World War I, his theorizing became more systematic, downplayed the fundamental role of the libidinal body, and became more focused on the structures and functions of representations that are "in the head." The work of some of Freud's early associates and successors, such as Carl Jung, Otto Rank, Sándor Ferenczi, and Wilhelm Reich, can also be characterized as the beginnings of a somatic psychology (and will be discussed in Chapters 6 and 7). In this sense, Freud's writings foreshadow the demise of what has been called the "modern episteme" (which will be discussed in Chapter 2), and anticipate the impulses of a postmodern era (Barratt, 1993). However, it can be argued that the twentieth century's elaborations of psychoanalysis, including Freud's later writings, distance themselves — both in the assumptions and terminology of their theorizing and in their clinical practices — from the lived experience of embodiment. This can be shown to be true of each of the five major strands of psychoanalytic orthodoxy: the structural-functional school, Kleinian psychoanalysis, the various object-relational (relational or interpersonalist) schools, Kohutian self-psychology, and Lacanian psychoanalysis.

Structural-functional psychoanalysis — sometimes better and more broadly known as "ego psychology" — has had an enormous influence on the conduct of psychotherapy, particularly in North America. Its major sources are in Freud's *The Ego and the Id*, which was written in 1923, and *Inhibitions, Symptoms and Anxiety*, which was written in 1926. Its most eminent elaborators (in addition to Freud's daughter, Anna, who wrote *The Ego and the Mechanisms of Defense* in 1936) were immigrant American psychoanalysts such as Heinz Hartmann, Ernst Kris and Rudolph Loewenstein (whose major writings were mostly undertaken in New York in the 1950s), as well as David Rapaport. Ego psychology was further popularized by men such as Jacob Arlow and Charles Brenner (through the 1960s and 1970s), and continuing into the 1990s with subsequent elaborations (by theorists such as Dale Boesky and others who are sometimes called "neo-structuralists").

The central theme of this version of psychoanalysis is the manner in which the organized aspect of the mind's functioning — the structure and defensive functioning of the ego — produces cognitions and emotions that govern our behaviors and that are compromises between the demands of three hypothetical forces. These forces are, (1) the internal drives, which impact the ego as the representations of "drive derivatives," (2) external reality, which is hypothetical in the sense that it is always "reality" as the ego construes it or represents it, and (3) the internalized structures of superego and ego-ideal, which are approximately the representations that generate the regulating forces of guilt and shame.

In a sophisticated manner, this version of psychoanalysis endorses Cartesian dualism (the separated mind regulates the body and its behaviors) and installs an analytic epistemology in the Kantian tradition (Barratt, 1984). This sidelines the experience of the body, and keeps therapy mostly "in the head." In Anna Freud's 1936 book, for example, the body is treated as a major challenge for the ego's managerial abilities, preoccupying several aspects of its various "lines of development." The marginalization of the body in ego psychology is somewhat obscured by the fact that structural-functional theorizing preserves the elder Freud's terminology of instinctual drives (which are held to have bodily origins) even while subordinating them to the ego's representational and defensive activities. The theoretical edifice thus renders the most grounding dimension of human experience into a sort of abstraction. Likewise the libidinal body, which as we will argue is the fount of all psychological experience and is recognized as such in Freud's early writings, disappears from structural-functional psychoanalysis. It is replaced by the analysis of incidental "sexual behaviors" and bodily activities that are under the governance of the ego. In short, ego psychology and structural-functional psychoanalysis treat the body not as a source of wisdom to which we might listen, but rather as a mute "thing" whose chaotic impulses have to be managerially governed by the organized structures of the mind. Kleinian psychoanalysis, which was the major rival to ego psychology from the early 1940s to the end of the twentieth century, treats the lived experience of the body in a similarly incidental manner — although this is only paradoxically so (Klein, 1921–1963). Although Melanie Klein always labeled her contributions as Freudian, her major papers in the 1940s and 1950s, together with the work of her London-based colleagues, gradually constituted a distinctive school of psychoanalysis. This progression had enormous influence on therapeutic practice through the twentieth century, especially in Europe and South America. Kleinian psychoanalysis and its later "neo-Kleinian" elaborations take their inspiration from the object-relations perspectives that Freud offered in his so-called metapsychological papers written between 1914 and 1918. Together with a somewhat aberrant reading of Freud's notion of *Totestrieb* (usually translated as "death drive"), which he presented in his 1920 *Beyond the Pleasure Principle*, these works formed the basis of Kleinian thinking.

The central theme of the Kleinian perspectives is the individual's struggle with primordial aggression (which Klein later reconceptualized as primal envy) and how this affects the individual's developmental progression from the paranoid-schizoid to depressive conditions of psychological functioning. In this depiction of mental life, Kleinians refer extensively to representations of the body. Paranoid-schizoid functioning is chaotic in that, before the individual's psyche becomes more depressively organized, it is troublesomely preoccupied by phantasms of breasts, wombs, penises, and the like. This is the mêlée from which our psychological life develops. But the point is that these breasts, wombs, penises, and the like, are *phantasies;* Klein deliberately uses this term to distinguish these representations from the more organized "fantasies" that the ego can entertain as narratives. That is, Kleinian psychoanalysis may appear interested in the body, and its discourse may refer extensively to the phantasied activity of body parts, but these references are representational, having only a tenuous and distanced relation to our lived experience of embodiment. This argument has been well articulated by Efron (1985) and others. It is in this sense — and here is the paradox — that despite all the "bodytalk," the body, as our embodied experience, actually disappears from Kleinian psychoanalysis to be replaced by the dynamics of all these phantasies *about* bodily experience.

There are many other versions of psychoanalysis that are, in one way or another, object-relational, relational, interpersonal, or self-psychological. The so-called independent school of object-relations developed largely from the late 1940s onwards as a series of efforts to meld ego psychological and Kleinian insights. Donald Winnicott, Michael Balint and Ronald Fairbairn, all of them practicing clinically in the United Kingdom, are among the best known of the earliest contributors, although by the 1960s the work of Margaret Mahler and subsequently many others, such as Otto Kernberg, became professionally popular. Later, as these perspectives became known in the United States, new brands of psychoanalysis developed — relational, interpersonal, intersubjectivist, and so on. Many of these were not only influenced by object-relational and ego-psychological perspectives but also by self-psychology.

The latter is largely attributed to the 1970s writings of Heinz Kohut (an immigrant psychiatrist who practiced in Chicago). Kohut started his career in the structural-functional tradition and developed a theory of what are called narcissistic transferences. This was gradually elaborated into a divergent and distinctive theoretical framework, which abandoned reference to the ego, and focused on the way in which the self maintains its sense of cohesion. The self is, essentially, a representational structure or set of structures, which maintains itself by the use of various "selfobjects" (Kohut dropped the customary hyphen between these words in order to make a theoretical point). Strikingly, terms such as "body" and "sexuality" are almost entirely absent from Kohut's collected works, and this epitomizes the extreme extent to which psychoanalysis distanced itself from bodily experience from World War I onwards.

Finally, the psychoanalytic interpretations of Jacques Lacan must be mentioned, since he proclaimed a "rereading" of psychoanalysis between the mid-1950s to the early 1970s that has subsequently become enormously influential especially in Europe and South America (Lacan, 1972, 1977). Lacan's writings and his famous *Séminaires* (many of which are still neither translated nor even released for general scrutiny) purport to herald a "return" to Freud's early discovery of the unconscious, which he reread in terms of Saussure's structural linguistics, and in a manner that is vehemently critical of both ego psychology and Kleinian psychoanalysis, as well as all subsequent versions. Lacan's theories are enormously complex and his writings deliberately allusive and obscure. However, for our purposes, what is important to grasp is his proposal that the unconscious is structured as a language (and it is, for Lacan, a transpersonal structure that could be mathematically specified). It is immutable and phallocentric, and is the "Other" (which Lacan capitalized in order to make a theoretical point) as far as the speaking subject is concerned. This Other produces all the operations articulated by the conscious subject. Thus the subject is never author of its own meaning. Rather, the experience of the human subject is constituted by this unconscious system of representational rules and transformative regulations. It follows that "experience" occurs in three "registers." The main register is that of the symbolic order (the "Symbolic") which is the transpersonal system of rules of signification within which and by which the speaking subject is constituted. The "Imaginary" (which has little or nothing to do with imagination in the ordinary sense) is the register of speculative dualities responsible for the ego's illusion of its own substantial existence. The "Real" (which has nothing to do with reality in the ordinary sense, since the experience of "reality" is furnished by the symbolic order) points to the unthinkable and unsayable rupture of the symbolic order, somewhat akin to the abyss of deathfulness. It would seem that none of these registers present the wisdom of bodily experience, and in this sense one of the major criticisms of Lacanian ideology is that it advances a sort of capsized Cartesianism (Barratt, 1984, 1993; Wilden 1968, 1972).

While Lacan's followers would understandably protest this brief caricature of the main coordinates of his theory, it serves our purpose in one important respect, for we can see that there is, in the rather abstract complexities of his theorizing, little or no respect for bodily experience. Despite all Lacan's talk about sexuality, and his theorizing of desire in terms of what he calls our *manque-à-être* (which approximately translates as "lack-of-being" or "want-of-being"), the grounding of the psyche in the lived experience of our embodiment is precluded from the Lacanian cosmology. It is left to influential feminist psychoanalysts, who critique Lacanian precepts, to attempt to remedy Lacan's disregard for the vivacity of bodily experience (here one thinks of the important work of Julia Kristeva, Luce Irigaray, Hélène Cixous, and others).

While these few pages can scarcely pretend to be a survey of the history of psychoanalytic thinking through the twentieth century, they are perhaps sufficient to illustrate and underscore one crucial point. The history of twentieth century psychoanalytic psychology has largely been a retreat away from Freud's seminal insights about the grounding of the psyche in the energetic experiences of our embodiment. At best, his insights have been reduced to a set of badly presented propositions about phases of childhood development — the oral, anal, phallic, latent, and genital stages that continue to be rehashed in every undergraduate textbook. These are propositions that almost entirely miss the wisdom of Freud's insights into libidinality — the erotic foundation of our psyche. At worst, Freud's later writings, and almost all of those of his successors, seem to suppress altogether his inchoate formulations of somatic psychology.

The progress of psychoanalysis through the twentieth century has directly or indirectly stimulated the entire field of psychotherapy, which developed dramatically throughout the greater part of the twentieth century. It diversified both in its theorizing and in its clinical methods, and it became ever more strongly established within the mental health industry as well as throughout the cultures of Europe and the Americas. However, this apparent progress was accompanied by an increasing neglect of the body. One benchmark of this is as follows. In the index of Freedheim's otherwise excellent anthology, *History of Psychotherapy* (1992), which surveys the twentieth century as "a century of change," the terms *body, sexuality* or *sex*, and *somatic* are nowhere to be found (and Wilhelm Reich is mentioned only in passing as a historical figure who was outcast from the psychoanalytic movement). This is, I believe, yet another interesting indicator of how much twentieth century psychology has endorsed and entrenched the alienation of mind and body.

Interestingly enough, in the very last years of the twentieth century, there was evidence of an inkling within the orthodoxies of organized psychoanalysis that sexuality — and indeed the libidinal body — needed to be remembered and theoretically reinstated or rehabilitated (Bloom, 2006). At the end of the 1990s, the Congress of the International Psychoanalytic Association in Barcelona was titled "Psychoanalysis and Sexuality," and in the same period the American Psychological Association's Division of Psychoanalysis directed some attention to this topic. Reading the proceedings of these conferences, on the one hand, one might be impressed by the importance of these papers as an effort to respond to the emergence of somatic psychology and bodymind therapy that was and is happening mostly outside the official organizations of psychoanalysis. On the other hand, one might read these proceedings and be struck by how remote the presentations often seem in relation to sexuality as a lived experience of our embodiment. Somewhere early in its history, psychoanalysis lost the wisdom of the body, spent decades retreating from this wisdom (and regressing theoretically to various versions of Cartesian dualism), and is only now trying to reconnect with its roots in somatic psychology (e.g., Blechner, 2009; Muller & Tillman, 2007).

Somatic psychology as the future of the discipline

Looking at the twentieth century history of psychology, in the way we have just done, one can tell that the discipline is at the crossroads — so too are the associated visions of psychotherapy. It is very probable that the discipline seems so disjointed precisely because it is at the beginning of a major transition that will affect the fundaments of our way of understanding the human condition. Old paradigms are crumbling; new ways of knowing and being are emerging. So let us now make a bold prediction in three parts:

- That by the middle to end of the twenty-first century, psychoanalysis as we know it today will no longer be much in evidence.
- That by the middle to end of the twenty-first century, cognitive behaviorism will no longer be credible as science — although, grievously, technologies of manipulation may continue to exist because, unless the geopolitical organization of society and its planetary cultures are revolutionized toward a liberatory vision of the processes of being human, these technologies will still be utilized by the dominant social order.
- That psychology will become *somatic psychology* and psychotherapy will be *bodymind therapy*.

And what is this discipline? It is not a psychology of representations "in the head" that might be referentially *about* matters of the body. Rather, we will define somatic psychology as follows:

Somatic psychology is the psychology *of* the body, the discipline that focuses on our living experience of embodiment as human beings and that recognizes this experience as the foundation and origination of all our experiential potential.

And we will define bodymind therapy as follows:

Bodymind therapy is healing practice that is grounded on the wisdom of the body and guided by the knowledge and the vision of somatic psychology.

The prediction that this discipline, in both its theories and its healing praxis, will be the wave of the future is neither far-fetched nor whimsical. This book presents supporting evidence that will show you why this is so.

2
Epistemic Shifting

The major intent of this book is generally to appraise the evidence that psychology and psychotherapy are currently articulated in the process of a major transition that will eventually affect the fundaments of our way of understanding the human condition, and specifically to comprehend the role of somatic psychology and bodymind therapy in facilitating this process of change. So in this chapter and the next, we will approach the issue of this change from two very different perspectives.

In this chapter, we will discuss the character of "epistemic" shifts — shifts that do not merely entail a circumscribed switch from one paradigm to another, but rather that reverberate through every aspect of our understanding of what it means to be human and of the nature of the universe in which we humans live. This discussion will be somewhat philosophical and historical.

Then in Chapter 3, we will focus very specifically on bodymind therapy, and we will illustrate schematically some of its healing practices. The purpose of this illustration will be both to offer readers who are entirely unfamiliar with such practices a rudimentary sense of what might be involved, and to demonstrate how different these clinical methods are from the twentieth century's prevailing notion that psychotherapy has to be almost exclusively a matter of *talking about* one's life and its eventualities.

On the notion of the episteme

We owe to Foucault (1966, 1969) and those who have followed his work the useful notion of an episteme (from the Greek, *épistémé*), as a way to understand some of the foundations of human culture, and specifically the shifts that western cultures have undertaken. An episteme is a "masterdiscourse" that governs the possibilities of expressing our experience — the possibilities of our thinking and speaking intelligibly. According to Foucault, all the various theories, epistemologies, and paradigms, by which we make sense of things, occur within, and are determined by, an underlying episteme. The episteme thus sets the limits and conditions of our capacity to understand our selves and our world. We cannot fully comprehend the episteme within which we operate — yet we cannot think outside its governance — because the episteme sets the ground for matters as fundamental as our experience of time and space, as well as identity and difference.

Foucault's methods, which are called an "archaeology" or "genealogy" of knowledge, suggest that each historical era of human culture is governed by an underlying episteme. Each episteme emerges from its predecessor, controls the possibilities of thought and action for several centuries, and eventually collapses. This notion not only enables us to describe how the modern episteme emerged from the medieval; it also enables us to understand that the significance of the twentieth century is the way in which it has brought us to the realization that we are currently in the midst of a process of epistemic collapse.

Let us examine this in a little more detail. In western cultures, the medieval era was characterized by what might be called "theocratico-theological reasoning" (which produced a metaphysical psychology). This episteme started to show evidence of strain perhaps as early as the twelfth century, but went into collapse definitively by the late fifteenth century. As the modern era emerged, with the rediscovery of Hellenic learning, from Hebraic and Islamic sources, there was a discernible drift away from the medieval mode of reasoning; for example, with the emergence of disjunctive knowledge and what has been described as an "identitarian masterdiscourse" (Barratt, 1993, pp. 207–223). This emergence is associated in science with the "revolutions" of Nicolaus Copernicus (1473–1543), Galileo Galilei (1564–1642), and Isaac Newton (1643–1727). Its philosophical standard-bearers are most prominently Francis Bacon (1561–1626) and René Descartes (1596–1650), although, from different perspectives, the works of Immanuel Kant (1724–1804) and Georg W. F. Hegel (1770–1831) can be viewed as the philosophical pinnacles of the modern era. By the late sixteenth or early seventeenth century, the modern era was fully established, and its episteme came to govern western culture, almost seamlessly — up until its implosion. The modern episteme began to show signs of strain in the late nineteenth century. Through the course of the twentieth century, it visibly and dramatically began to break up and fall apart. A discernible process of epistemic collapse became evident. Now at the beginning of the twenty-first century, we are still in the midst of experiencing this profound shift.

The modern episteme is characterized by what might be called "analytico-referential reasoning," as has been well discussed by Timothy Reiss and many other scholars (Reiss, 1982, 1988, 2002b). In the modern era, language's rhetorical and logical function is assumed to guide reasoning such that the subject's reiterative construction of representations allegedly comes to mirror accurately the structural organization of the world that is external to, or other than, the subject's thinking. Subject and object function as each other's *other*. In this era, the world is assumed to be singular and understandable via the linear narration of "cause and effect." Doctrines or theories about the world can be empirically validated, and science is regularized by geometric and algebraic propositionality that corresponds to the immutable laws of nature (time and space). That is, the universe or nature as *other* than the thinking man (I use this gender deliberately) is assumed to be understandable to the human faculties of reason because it operates on an unchangeable lawfulness that is reflected in the logical and rhetorical functions of the analytico-referential masterdiscourse.

For our purposes what is salient here is the way in which *dominative mastery over the other* becomes the motif of the modern era — characterizing its epistemology, its ethics, and its ontology. The relations of knowing and being, as well as the precepts of moral propriety, all come under the aegis of domination. Human enterprises, both of reasoning and of commerce, confront a variety of "others." Nature is other than "man" and lies mutely "out there," waiting for him to penetrate her secrets scientifically and to plunder her resources (Bacon's texts express these sentiments in all their rawness). Women are other. They are the weaker sex, ruled disruptively by their passions, and in need of governance by the stable rationality of men (Descartes' texts are quite explicit on these tenets). The third-world and people of color are other. Their labor force and their resources await domination and exploitation by capitalist conquistadors; their souls await the blessings of missionary outreach and the promulgation of "civilized" values. Those who are not white, western and male, are "primitive races" who should feel blessed to be colonized, enslaved and exploited, by European and later American imperialist powers. These are the powers of those who assume that their mastery of science, their industrial prowess, and their Christian theology render them the "natural superiors" destined to control non-European peoples. Within the western world, ruling-class elites, with their capital resources and their access to education, treat themselves as self-evidently superior to their other — the agrarian peasantry and the industrial proletariat. Finally — and most pertinently for our purposes — the body is treated as other to the mind. As is well known and as was mentioned in Chapter 1, this is most famously formulated by Descartes. The mind is held to be inherently different from the body and, in a prescriptive frame of reference, the mind should be the agent in control of the "other" that is its own embodiment.

In the modern era, truth is supposedly demonstrated in the practices of mastery by domination — the prediction, manipulation and control, of the object of knowledge and conquest. That is, if you have dominative mastery over the other, then you can be presumed to "know" its secrets. Modern reason is identitarian and disjunctive, forcefully inscribing both patriarchal values and what has been called the "metaphysics of presence" (cf., Reiss, 1982, 1988). For example, the contribution of absence to the process of something being seen as present becomes hidden from consideration; an entity cannot both be and not-be. Women are defined in terms of whatever is not-man; children are viewed as deficient adults; and so forth. Indeed, what is crucial to grasp here is the way in which differences (the difference between the one and its other) are always construed in terms of domination and deficiency-defect. This is the logic of domination and subordination-subjugation: rational man over nature, over women, over children; rapine heterosexuality over the sensuality of all that is queer; white man over people of color; rich over poor; and master over slave. In the modern era, all this is understood as the "natural order of things." And central to all this is the assumption that the referential and rational representations of the mind — the conceptual, propositional or calculating operations of mental activity — are different from, and should preside over, the mute impulsiveness of the body.

In sum, immersed in the western culture of European and North American domination, we have all inherited a mindset for understanding the otherness of other peoples and cultures — as well as the illusory otherness of the body — that is at once hubristic and hegemonic. The one treats its other as subordinately different and ready for subjugation. The one often claims to be the beneficent purveyor of "civilization," of missionary values and the like, but its actual treatment of the other is almost invariably oppressive, repressive, penetrative, exploitive, and ultimately barbaric. The other is different and therefore held deficient; the apparent deficiency of the other supposedly constitutes an invitation to domination. In the modern era, the standards by which all things are to be assessed are always those of the dominant social and cultural order — white, male, Christian, affluent, aggressively ambitious in its own self-interests, and yet the bearer of values that are supposedly "civilized" and "civilizing."

The episteme of the modern era has deeply entrenched western cultures (and indirectly, because of the west's colonialist ambitions, the cultures of most of the entire globe) in at least seven tragic *"isms."* These are classism, elitism, racism (or ethnocentrism), gender-sexism, heterosexism, ageism, and ecocidalism. To these we must add that the modern episteme profoundly inscribed for all of us the conditions of our alienation from the living experience of our embodiment. And perhaps this alienation is indeed central to, and the wellspring of, all the other tragedies — because rendering our own body as other makes possible a mindset that treats the earth and all its inhabitants as other than our selves.

Charting the collapse of the modern episteme

What we have just briefly sketched as some of the salient features of the modern episteme has held sway over western culture for at least four hundred years, and persists to the present day. As yet, we cannot think and speak outside the limits and conditions of the modern epis-teme — the identitarian masterdiscourse. However, we have powerful intimations that this episteme is imploding around us, even while we are operating within it. These intimations have steadily intensified through the course of the twentieth century, to the point that we now know that our old ways of thinking are exhausted — ontologically, epistemologically, and ethically — but we do not yet know fully how to think *otherwise* (Barratt, 1993).

Let us now examine in a little more detail this notion that we are living in the midst of an epistemic implosion; sketching some of the changes that have occurred and are occurring in philosophy, and then mentioning some of the shifts that are happening in science.

Many commentators have venerated Hegel's philosophy as articulating the pinnacle of the modern era. Written in 1805, his *Phenomenology of Spirit* presented a vision of the universe as a progression in which everything could, and ultimately would, be comprehended — terminally enfolded into the immutable summation that he termed "absolute knowledge" (Hegel, 1977). The *Phenomenology*, together with Hegel's logic and his various encyclopedic efforts, claimed to represent the goal and completion of modern thinking, subsuming its major predecessors such as Kant's critical philosophy, Johann Fichte's practical idealism, Friedrich Jacobi's intuitionism, and Friedrich von Schelling's aestheticism. The Hegelian grand synthesis left subsequent thinkers of the nineteenth century challenged to demonstrate whatever might have been overlooked or omitted from Hegel's articulation of the character and conditions of knowledge. For Søren Kierkegaard (1813–1855), this concerned the existent individual and the vicissitudes of living experience (considerations that are pertinent to the emergence of somatic psychology, as we shall see). For Karl Marx (1818–1883), this concerned the material substrate and genesis of ideas in terms of the concrete relations of production (and, we might add, the grounding of ideation in the material substrate of the bodies involved in the relations of production). For Friedrich Nietzsche (1844–1900), this concerned what might be characterized as the ethics and aesthetics of existence (and some of his contemplations on the earth, the body, and the temporality of recurrence remain freshly relevant to the contemporary shifting of our perspectives).

These critiques of Hegelian philosophy all point toward the subversion of various aspects of the modern episteme. However, there is an additional feature of Hegel's work that bears directly on our concerns, namely his investigations of consciousness and self-consciousness in the early sections of his *Phenomenology*. For in these essays, it may be argued that Hegel demonstrates the limitations of what he calls "sense-certainty," the way in which what appears immediate is always mediated, and the dependence of the present "now" on what is absently then or there. The intimated subversion of the priority of presence — and the suggestion that the "now" is what we have but that it is always, so to speak, essentially empty — resonates in a strange way with Buddhist epistemological doctrines (Arnold, 2008; Bhatt & Mehrotra, 2000; Watson, 1998; Yao, 2005). It is also powerfully significant for our understanding of "holistic interdependence" that animates much contemporary work in somatic psychology. Hegel's philosophical writings, although they may be considered the glorious apotheosis of the modern era, are also a rich resource of ideas that point to the subversion of its episteme.

In a different vein, the work of Sigmund Freud (1856–1939) can be seen as both an elaboration and a radical critique of the Hegelian enterprise. Freud's threefold understanding (1) of the grounding of the human spirit in its libidinality, (2) of the embedded connectivity between individual and culture, and (3) of consciousness being formed as the dynamic and disguised returning of what it has itself repressed (the so called "discovery of the unconscious"), all contribute to the impending collapse of the modern episteme. As discussed elsewhere, Freud's writings, especially after 1914, can be characterized as a conservative defense of many of the assumptions of the modern era about the nature of humans and their place in the universe. However, especially between about 1895 and 1914, his writings also herald the ending of the modern episteme (Barratt, 1993).

Philosophically, the metaphysics of the modern episteme promulgated the unity, identity and immediacy of the thinking subject (which is the essential instantiation of our representational capacities) and the notion of the absolute (as the identity of the totality of all things, and as the enclosure of representational time), and these in turn specified all the possible relations of difference between the "one" and its "other." As has just been suggested, psychoanalysis subverts this metaphysics by suggesting that human consciousness never conforms to such specifications.

Other developments in the twentieth century also begin to erode the assumptions of the modern era. For example, Freud's contemporary, Edmund Husserl (1859–1938) inaugurated phenomenology as a philosophical method or discipline by which the prerogatives of the Cartesian subject might be re-established and our understanding of this subject deepened. Yet one of his most important and overlooked early works, *The Phenomenology of Internal Time-Consciousness*, almost entirely sabotages its own intent by showing how fragile or insubstantial is the experience of "now," and how dependent it is on our representational experience of whatever comes before or after it (Husserl, 1964). Similarly, Husserl's earlier inquiries intimated the problems encountered by any philosophical enterprise that attempts to secure unassailable foundations for logic (1970). It is telling that, by the end of his life in 1935, Husserl was incisively declaring the sciences to be in crisis (Husserl, 1974). All this will be discussed further in Chapter 8.

Although he did not employ the notion of epistemic shifts, Husserl's student, Martin Heidegger (1889–1976) promoted his own philosophical writings as standing at the closing of one epistemic era and at the threshold of something different. Although his most famous work, *Being and Time*, could be seen to have phenomenological underpinnings, in the course of his career Heidegger shifted toward what is more aptly characterized as "hermeneutic ontology." This is exemplified by works such as *On Time and Being* (1972), *On the Way to Language* (1982), or *Poetry, Language, Thought* (2001). This approach not only exposed the problem of thinking about being in terms of presence (a metaphysics that had entranced western philosophy since Socrates), but also called for a new approach to questioning our being in terms of the way in which being itself is housed within language. Despite Heidegger's alleged proclivities toward Nazism in his early career, his later philosophy came to have pronounced Buddhist and Taoist connections. As we will see later, Heidegger's work extensively influenced the emergence of poststructuralist thinking and even "postmodern" impulses.

Somewhat prior to Heidegger, and contemporaneously with Freud, the work of Friedrich Gottlob Frege (1848–1925) attempted to establish the foundations of mathematical logic and analytic philosophy, and had a far-reaching influence on Anglo-American philosophy through the twentieth century (appealing to the logical positivists, empiricist analytic thinkers, and the unified science movement). Most interestingly, although Frege resorted to a theory of truth that emphasized self-sufficiency and internal logical coherence, his writings brought modern epistemology to the brink of its crisis. They suggested that logic could never adequately prove the referentiality of our representations (implying, for example, that "facts" are always already mediated, and thus are always "deeply theory laden").

By the end of the nineteenth century, the assumptions about "experimental truth" that had guided the sciences of the modern era were coming under attack — even if this was an unintended or indirect attack, as in the case of the work of brilliant mathematicians and philosophers such as George Boole (1815–1864), Ernst Schröder (1841–1902), and Alfred North Whitehead (1861–1947). Efforts to establish the priority and the immutably mathematical foundations of logic, such as Husserl's early work, Whitehead's *Principia Mathematica* (coauthored with Bertrand Russell and published between 1910 and 1913), and Ludwig Wittgenstein's *Tractatus Logico-Philosophicus* (first published in 1921), were heroic, brilliant, and influential works, that were read avidly through the course of the twentieth century. But in a certain sense, they are all notable for exposing the limitations of their own enterprise. In the course of the twentieth century, confidence in the logical foundations of truth began to fall apart. Philosophy and the social sciences all undertook what has been called the "turn to language" (and away from the pursuit of mathematical models and the foundations of logic). The "cognitive turn" in psychology might be considered an aspect of this general turn toward language.

This "turn" was not only indicated by the shift in Husserl's thinking, by Wittgenstein's shift from logic to language (in his posthumous *Philosophical Investigations*), and by related developments (such as Heidegger's hermeneutic ontology). It was also exemplified both by the European advent of structuralism as inaugurated by the famous "Course in General Linguistics" taught between 1906 and 1913 by Ferdinand de Saussure (1857–1913), and by the North American advent of semiotics or the "science of signs" as inaugurated by Charles Sanders Peirce (which he started in the late 1860s and worked on until his death in 1914). We will return to these developments briefly in Chapter 8, when some of the developments in the human sciences that have influenced the emergence of somatic psychology are further discussed.

Alongside all these indications in philosophy that suggested potential ruptures in the modern episteme, twentieth century developments in the "hard sciences" definitively exposed the limitations of the modern masterdiscourse. Following the four revolutionary papers that Albert Einstein published in 1905 (on photoelectric effects and Max Planck's quantum theory, on Brownian motion and atomic theory, on electrodynamics and the radical theory of special relativity, and on the theory of mass-energy equivalence), our understanding of the universe has gradually been turned upside-down. The modern era's dichotomies of substance versus radiation, particle versus wave, and even determinism versus indeterminism, have all been shattered. Counter-intuitive experimental findings are in the process of shifting our comprehension of the world from a deterministic Newtonian and Maxwellian universe, grounded in principles of Pythagorean mathematics, Euclidean geometry, and Archimedean measures, to new modes of comprehension that were previously unthinkable.

With the advent of the "new sciences" — particularly at the subatomic level of quantum mechanics and the cosmological level of astrophysics — we have lost our grip on reality, so to speak. The general progress of science through the twentieth century certainly appears dramatic. This is largely due to the riptide of technological advances, as well as the shift from accomplishments subsequent to the industrial revolution to the accomplishments of computerized instrumentalization and the informational revolution. However, our confidence in the potential of technological innovation to save humanity from its own self-destructiveness is perhaps waning. It has been profoundly challenged by the growing realization that so many technological advances since the industrial revolution have had disastrous consequences for humanity and for the planet. The apparent progress of science is paralleled by an increasing insecurity not only about the ethical and humanitarian impact of technological accomplishments which are so readily available for misuse, but also about the actual nature of the universe that is opening itself to us. Contrary to the modern era, reality no longer seems external and monolithic. Flux prevails, and whatever lawfulness there is in the universe no longer corresponds to the identitarian metaphysics of the modern episteme. For example, time is no longer the linear, equable flow which Newton described as the assured foundation of our capacity to comprehend the world in which we live.

The new sciences are proving to us that the modern era's values of scientific distance and detachment, of depth and essentialism, of the technocratic imperative, and the masculinist notion of truth as mastery by domination, are all crumbling. A universe of interdependence — foretold in Vedic, Buddhist, Taoist, and many indigenous teachings — is now being demonstrated scientifically. The dominative separation of subject and object, and along with it the dichotomies of man's mind over nature, mind over body, and so forth, are proving illusory as the necessity of thinking in terms of nonlinear and dynamically complex systems is pressing itself upon us (cf., Cowan, Pines & Meltzer, 1999; Kauffman, 1996, 2002, 2008; Morin, 2008).

The emergence of the postmodern episteme

Although it would be foolishly premature to write about the postmodern episteme (despite the fact that the term "postmodernism" has already become popular parlance), as we work and play at the beginning of the twenty-first century, we are beginning to become acutely aware of the need to think, speak, and act, *otherwise* than the ways of thinking, speaking, and acting that were installed within the modern era. The coordinates of the modern episteme are proving themselves unstable. They are in the process of shifting radically. We cannot know how different human life will be in the course of the next fifty or more years, but we do know that it will necessarily be otherwise than it is now.

We have pointed to some of the philosophical and scientific transitions that intimate the ending of the modern era. There is also, as might be expected, a cultural, political and socioeconomic context to these changes. It is important to consider this because one of the myths of the modern era is that science and philosophy progress autonomously. However, the modern era's emergence was fueled by the rise of capitalism, the industrial revolution and the colonial-imperialist expansion of market economies. Each of these events was structured to make the European ruling-class (and later the USA's elite) accumulate wealth at the expense of the various others that it came to subjugate (women, children, the agrarian peasantry, the industrial proletariat, people of color, and the third-world). Through the twentieth century, the modern episteme retained its ideological grip by means of its technological advances, even while its underlying intelligibility was rapidly becoming compromised. However, the twentieth century also testifies to other modes of transition, as capitalism becomes transnational and as cultures become globalized.

The twentieth century witnessed the shifting of capitalist structures from national to transnational, and of political structures from those of colonial-imperialism to those of post-colonial imperialism. The robber barons and later the captains of industry have been replaced by the anonymous boardrooms of transnational capitalism. The twentieth century witnessed the globalization and homogenization of planetary cultures following the information revolution — to the point where almost the entire world has access to the Internet, as well as exposure to the ideologies promulgated by the American media. The crass commercialism and platitudinous attitudes conveyed through television sets and computer monitors, along with the sinister political dealings of the military and paramilitary organizations that serve the interests of corporate boardrooms, are now everywhere. These developments have entailed the dominative spread of European and North American economic structures and cultural values across the globe (Baylis, Smith & Owens, 2008; Reiss, 2002a). However, against the momentum of this expansion, the twentieth century also witnessed the emergence of new freedom movements acting against the oppression of people by race, gender, sexual orientation, age, and so forth. It is crucial to recognize that such movements have also brought with them new, and newly marginalized, ethical sensitivities.

At least in some quarters of the western world, there is a dawning sensitivity that *difference* might have inherent value, that it might be an occasion for the celebration of otherness, an occasion for growth and wisdom. This sensitivity has taken us to the verge of a realization that difference might be ethically configured otherwise. Although acting against the political mainstream, an increasing number of peoples from all walks of life are beginning to insist that differences should no longer be constituted in terms of the invocation of ideas about deficiency, nor should they be construed as an invitation for the politics of domination (Ashcroft, 2008).

Within psychology, a critical movement has emerged in recent decades (cf., Fox, Prilleltensky & Austin, 2009; Parker & Spears, 1996; Prilleltensky & Nelson, 2002; Sloan, 1996, 2000; Teo, 2005; Tolman,

1994). This is a multifaceted critique of the mainstream's collusion with the western imperialism, with the wealthy, the male, the white, and the dominant order. It involves critical perspectives that are variously anti-capitalist and anti-imperialist, following the work of such eminent Marxist psychologists as Lev Vygotsky, Georges Politzer, Lucien Sève, and Klaus Holzkamp (e.g., Brown, 1974; Holzkamp, 1972, 1992; Lethbridge, 1991; Sève & McGreal, 1980). There is also important feminist criticism of the patriarchal and masculinist ideologies that have governed psychological science (e.g., Benhabib, Butler, Cornell & Fraser, 1995; Gergen & Davis, 1997; Griffin, 1978; Keller, 1985; Merchant, 1980; Reinharz, 1992; Spivak, 1999; Winston, 2004). There is postcolonial criticism of the ethnocentric and racist agendas that have subtly — and not so subtly — structured the development and implementation of this science (e.g., Ernst & Harris, 1999; Fay, 1996; Guthrie, 1998; Harding, 1998; Howitt & Owusu-Bempah, 1994; Memmi, 2000; Mills, 1997; Richards, 1997; Winston, 2004). And there is a body of what might be called postmodern criticism, which is variously directed toward the instability of the subject, the insufficiency of the absolute, the subversion of temporality, and the deconstruction of historicist narration (e.g., Barratt, 1993; Kvale, 1992; Natoli & Hutcheon, 1993; Rose, 1996; Rosenau, 1992). We will return to a discussion of these emancipative psychologies in Chapter 15.

It is a tragic irony that this dawning of a postcolonial or postimperialist consciousness (class consciousness, ethnic consciousness, indigenous consciousness, feminist consciousness, queer consciousness, and liberation consciousness of all sorts) occurs during a period in which indigenous cultures are vanishing or, more precisely, are being extinguished, at an unprecedented rate (Barnard, 2002; Maybury-Lewis, 1992, 2001). In parallel, the planet is being brought to the brink of toxic suffocation (Broswimmer, 2002; Kolbert, 2006; Lovelock, 2007). Ironically, just as we seek different ways of thinking about differences, we find ourselves living in a technological world in which the availability of electronic media is culturally homogenizing the entire planet (Appadurai, 2001; Jameson, 1998; Nettle & Romaine, 2000).

It is additionally ironic that, with the dawning of this emancipative sensitivity, comes quite recently a resurgence of vicious fundamentalisms around the world involving all three Abrahamic "religions of the book" (Jewish, Christian, Islamic) as well as other modes of evangelical dogma (Almond, Appleby & Sivan, 2003; Ruthven, 2007). The hallmark of a fundamentalist belief system is, of course, the conviction that there is one right way to think and act; fundamentalisms thus stand for the erasure of whatever is "other."

In short, the barbaric ideologies of the modern era may have lost even the appearance of an underlying philosophical and scientific intelligibility, but they are far from fading away quietly. Technology and political self-interest keep people subscribed to the masterdiscourse of domination. This continues even while philosophy and science (as well as considerations of the ethicality of humane caretaking and planetary stewardship) tell us very clearly that otherwise ways of aligning ourselves with the universe need to emerge and are indeed emerging.

The potential for *an otherwise notion of otherness* is crucial to the emergence of new ways of thinking, speaking and acting — crucial to the possibility of the emergence of an epistemic era following that of the modern. We know that the old dichotomies — mind/matter, wave/particle, radiation/substance, subject/object, me/not-me, rationality/irrationality, indeterminism/determinism, male/female — no longer serve us, no longer reflecting what we know of reality. But we do not yet quite know how to think, speak and act, otherwise. In this respect, there are at least three conspicuous challenges that we are in the process of addressing (even while not yet knowing quite how to address):

- *Interconnectedness:* We now know that things are not separate in the way that we customarily construe them to be. We know that we must learn to think and act otherwise in relation to the differences that appear between the "one" and its "other(s)." We know that all matters are inseparable, and yet we cannot simply deny differences. But we do not know quite what to do with this knowledge. On a metaphysical and scientific level, we now know that nothing is really separate and autonomous, but that all the entities and events of the universe are dynamically interdependent — through the past and into the future. On an experimental and philosophical level, we know that the objects of our knowledge transform themselves in response to the operations of knowing. On a sociocultural and political level, we know that the wealth of the rich depends on the labors of the poor; the dominance of the one depends on the exploitation or erasure of the other, whether the "other" is a matter of class, color, gender, sexuality, age, or belief. On the most mundane ecological level, we now know that one cannot dump trash in an "other" part of the planet, without the act of dumping sooner or later affecting every part of the planet. We know the simple truth that the planet is abundant, but not infinitely so. While we now know all these things, we do not quite know how to think and act in relation to them. We only know that the precepts of the modern episteme have failed us.

Temporalities: We also now know that, although time may seem to move like the singular, straight line of an arrow's flight (which was Newton's analogy), it may actually move in ways that are multiple, wavelike, curved or circular. Ideas that seemed bizarre in the late 1800s and early 1900s now seem matters of the utmost seriousness. An example would be Nietzsche's conjectures about eternal recurrence. Another example would be Freud's speculations about the "timelessness" of the repressed, and about the constitution of consciousness by the repetition compulsion and by what has subsequently been called the "narratological imperative" (Barratt, 1993; Wood, 2001, 2007). A further example would, of course, be Einstein's and later Max Planck's vision of relativities, which initiated entirely new and as yet unfinished ways of thinking about the relations of spatiotemporality. We now know that the relations between moments of instantiation may appear linear and unitary, but are actually nonlinear, dynamic and complex, just as we know that matters of apparent substance dissolve on scrutiny to reveal everything to be a dance of energy. We now know that the phenomenological experience of "now" is meaningless in terms of the scientific notion of time as a sequencing of energy dispersals. Yet we also know that our narratological constructions of whatever has been actual in the past, and our expectations or fantasies about what is figurative or in the future, are all merely *re*-presentational. We know that experientially the present moment, the "I-now-is," is all we have (so to speak), but we also know, from Buddhist insight to contemporary deconstruction, that this experience is actually that of emptiness, that it is an experience supported by the processes of absence (Glass, 1995; Park, 2006; Wang, 2001). All this impels us toward the necessity of new ways of thinking and acting.

Ethicality: The notion of Heraclitus and other pre-Socratics — that change is somehow a more fundamental feature of reality than are the entities that undergo change — begins to make sense today, even though it has previously been condemned to senselessness. In our modern mindset, we insist that things must exist before they can change; in our contemporary mindset, we begin to question even this. Yet buried in the complexities of his writing, Freud suggested that the mind discriminates quality before it discriminates existence (Forrester, 1991). Moreover, although this may even now seem like a far-fetched allegory, contemporary physics suggests that the quality of a material event is determined before the matter comes into existence. Gradually through the course of the twentieth century, we have come to understand that the priorities of epistemology and ethics must be reversed. In the modern episteme, epistemology (our knowledge whether something exists or not) was assumed to be anterior to, and to take priority over, ethics (our wisdom as to how matters are to be treated). This accounts for the modern era's consistent confusion of ethicality with the promulgation of moralizing codes. With the collapse of this episteme, ethicality comes to take priority over epistemology — how we treat matters comes to be more significant to us than debate over the existence of these matters. This is well articulated in the philosophical writings of Emmanuel Levinas (Levinas, 1969, 1990, 1998; Levinas & Cohen, 1990; Vries, 2005), and features directly or indirectly in much of the current work on liberation philosophy and critical pedagogy (McLaren, 1994; McLaren & Kincheloe, 2007). This sort of ethical essentialism is inherent in Dharmic and Taoic spirituality (contrasting them, for the moment, with the Abrahamic traditions that have been so influential in the promulgation of moralizing codes). We know that ways of living ethically must be engaged, and that this task is far more important than the accumulation of further knowledge on the level of factuality and technology. But as yet, we may still be unclear as to how to proceed with this mandate.

In the light of these considerations, it is surely evident that so much of twentieth century psychology developed within the death throes of a dying episteme. It is surely evident that the paradigms of behaviorism, cognitive-behaviorism, and most of the schools of psychoanalysis, are concordant with the designs of the modern masterdiscourse. And it is surely arguable that the science of the *psyche*, along with its healing practices, will — as the modern episteme gives way to different discourses — approach the wisdom of our embodiment in an entirely different manner. This then is the context of deep epistemic shifting within which the emergence of somatic psychology and bodymind therapy are to be articulated.

Illustrations of Bodymind Therapy

We stated in Chapter 1 that somatic psychology is the psychology
the body (as distinct from a psychology *about* the body, or an enterprise that directs its activities *at* the body). The prepositional distinction is perhaps profoundly significant, and points to a radical shift
in orientation. This shift from *about the body* or *at the body* to *of the
body* heralds a difference in the spatiotemporal or ontological relations,
as well as the ethical underpinnings, that are engaged within the discipline of psychology — the discipline that is responsible for inquiring
upon and healing the human *psyche* or soul. In this chapter, we will
offer just four vignettes to illustrate the sort of healing practices that
comprise this applied dimension of somatic psychology. Of course,
many more vignettes could be offered as illustrative of the methods of
somatic psychology; however, since the intent of this book is mostly to
establish the emergence of this discipline, a proliferation of examples
might well distract from our primary purpose.

In the outdated climate of a science that values only "evidence-
based" findings that are externally observable, measurable, and appear
to be the result of unilateral manipulation, the practice of illustrating
truthfulness by anecdote perhaps needs to be briefly defended. Although
vignettes may not meet these narrow standards of evidence, they are
necessary in order to offer interested parties some sense of what is involved
in processes that are neither public nor readily measurable. For better
or for worse, almost the entire history of contemporary psychotherapy
— from Freud, Alfred Adler and Carl Jung onwards — has run on the
practice of responsible anecdote. The notion of responsibility at issue
here is, of course, a matter of considerable debate. Some vignettes are
unconvincing and, like any literary text, open to critique. Others fas-
cinate and compel. But all are inadequate portrayals of the complex

processes to which they refer, and none are to be taken as "proof" of a specific method's effectiveness. These cautions apply to these four vignettes. They are *real*, although partially fictionalized for the sake of confidentiality and ease of presentation. They may even be "typical" or mundane. But they are not, and could not be, "the whole story" — as if such a thing is ever possible. Rather, they will, hopefully, serve to illustrate the distinctiveness of the bodymind approach to psychotherapy.

• *Vignette A:* Lara sits facing her therapist, as she has twice weekly for several months. An attractive dark-skinned woman in her early twenties, she lives alone and came to treatment because she longs for partnership and motherhood, but has never been involved in an adult relationship with anyone beyond casual acquaintanceship. Men are erotically interesting to her. Yet, in her therapy she has come to recognize that she has been employing subtle but effective ways to keep them at a distance. As a physician employed by a major teaching hospital, she enjoys the company of her many female coworkers, but is not sexually attracted to any of them and is puzzled that deeper friendships with women rarely seem to develop. The therapist experiences Lara as "somewhat closed down," but likeable, articulate and very intelligent. During the early weeks of treatment, Lara led the dialogue by relating her life story enthusiastically and in some detail. Yet in recent sessions it already seems as if she is like a performance artist who has exhausted her material.

Lara sits silently. "I don't know where to go next. I've told you so much, and you've been comforting as you listened. You've also been helpful. I can see so much better how I keep men away from me, and previously I didn't even see myself doing it. I'm also more aware of being strangely fearful of them, even the ones who are clearly nice. But this session, I feel like I have nothing more to say, which is silly because there must be so much more." Again, she falls silent.

The therapist invites her to take a deep breath, close her eyes if she wishes, and scan her mind and her body to see if she notices anything arising. Lara, eyes closed, breathes deeply and softly at first and remains silent, but after a few moments the breathing seems slightly shorter and shallower. She says, "I feel tightness in my upper chest, my neck and my jaw." The therapist says, "I invite you to breathe into that tightness, and see what happens."

Lara begins to well with tears, and she gulps as if for air. "I adored my older brother. We used to take naps together when we were little, and we always shared a bedroom because my parents' house was so small. My brother was always very moral and mild-mannered, a great student, and I admired him. He was two years older, and used to tease me affectionately almost all the time. What just came to mind was one night when I was eleven years old. I awoke to find him standing over my bed. He had pulled down the covers and was looking at my budding breasts ... I don't know, I can't remember clearly, but I think he may have been naked. I had totally forgotten about this until just now."

What is illustrated in this brief vignette is the way in which a simple act of attending to the messages of our embodiment can seem to release a memory into awareness. In this case, the "message" of the body involved tightness in the upper chest, the neck and the jaw. In terms of subtle energy systems, this is approximately the region of the fifth chakra. In anatomical terms, it is an area of considerable muscular-skeletal complexity. Lara's tightness is accompanied by a transition into a breathing pattern associated with anxiety. Attending to these as messages, in a rudimentary version of what Eugene Gendlin (1982, 1991, 1997, 1998) discusses as "focusing," served to bring into awareness a crucial, but strangely forgotten, image and memory (Cornell, 2005; Weiser, 1996).

For the purposes of this illustration, we do not need to go much further with Lara's personal journey, except to note that the unfolding of subsequent memories gradually tarnished the image of the brother's entirely "moral and mild-mannered" character. It also brought forth several episodes in which Lara had fellated his adolescent penis. What is more important to note here about the retrieval and reintegration of these memories (which were often confused as to whether the fellatio had been willingly volunteered, affectionately coerced, or both) is that the recollected content was invariably preceded or accompanied by vivid messages from her bodily experience.

The messages had two aspects. On the one hand, they had a general implication. The constriction of the throat area communicated Lara's conflict over the energetic surfacing of intense emotions and her ability to "speak out" about them. These emotions included shame and guilt over the recognition that she may have been a willing participant in the fellatio, and may to some extent have enjoyed her brother's attentions. They included the recollection of fear, as well as some sexual excitement over the memories that had surfaced; for example, after one session she self-pleasured thinking about her brother's penis in her mouth. And this mixture of emotions included yet more shame and guilt accompanying the idea that she must be a "bad person" to have these pleasurable feelings about an incestuous act. On the other hand, the constriction of the throat area had a more specific symbolic significance. It reenacted the threefold anxieties of an eleven year girl. She had been fearful of the size of her brother's penis, which instead of "holding still like a lollipop" occasionally thrust toward the back of her throat. She had always been afraid of gagging and choking. She was also fearful because of her childhood fantasy that perhaps impregnation could occur orally (as an intelligent pubescent girl, she knew better, but was not entirely convinced). And she was additionally fearful because she now had a secret, which she could never speak out, until the healing work being accomplished in the therapy.

• *Vignette B:* Jake chooses to lie on his therapist's couch. He is tired. Nearing his sixtieth birthday and the pleasant prospects of retirement to a warmer climate, he is upset over his company's recent reorganization. He has been required to transfer from a division in which he had worked for many years — successfully and with significant responsibilities — to another department, in which he is given less work, less responsibility, and has to suffer what he considers the "indignity" of reporting to a much younger man. The transfer is a "promotion" with greater salary, but it has hurt his pride. Jake's financial situation is such that he could not decline the position, even though he would have liked to, because he has decided that he needs just a few more years of employment before he and his gay partner have enough money to implement their retirement dreams.

Jake's therapist experiences him as very soft spoken and sad. He seems to lack much energy for life. He loves his partner, but is rarely sexual with him or by himself. He spends much of his spare time in rather passive pursuits such as watching television, and he seems to have little enthusiasm for anything. Even a discussion of his cherished plans to retire to a warmer climate in a gay-friendly community elicits only a slight increase in his animation. Tears frequently trickle down his cheeks, as he repeats the stories surrounding his anguish over the company's treatment of his position.

The therapist wishes to feel genuine empathy for Jake's plight, and indeed he does feel sad for this man's obvious sadness. But the therapist also finds himself feeling rather irritated with Jake's rather passive and self-pitying tone; the seemingly endless repetitions of the same stories are difficult to attend to caringly.

Jake often goes over his material as if it were frequently rehearsed. He dislikes the young supervisor, finding him arrogant and demeaning.

Jake is bored by his work because he has years of professional experience and an expertise that far exceeds what the job now requires. He suspects that the young boss dislikes him, and he speculates that the man may be homophobic. The younger man has a conspicuously active hetero dating life, and Jake admits to himself that he has no valid evidence for the allegation of his supervisor's homophobia. Jake frequently feels that he is being demeaned, but quickly realizes that this is just a feeling. In actuality, this rather meticulous supervisor is just "all business, business, business" — nothing is personal.

In one session, Jake adds a detail to his stories. He gets "gut aches" on his way to work. He has tried varying his breakfast menu, but almost invariably his stomach tightens as he makes the drive to the office. The achiness dissipates as the day wears on, but it seems strangely anomalous. Jake has had regular physical examinations, and has always been told he is in excellent health. His diet is nutritious and his intake modest; his bowel movements are regular and well-formed.

On one previous occasion, the therapist had invited Jake to try an adapted method of "body dialogue" (cf., Griffith & Griffith, 1994; Osho, 2005; Rous, 2006; Stone & Stone, 1998). The effort was unsuccessful. Jake tried the procedure in a rather passive and compliant manner, later deriding the method as "feeling a bit silly." Despite this unpromising start, the therapist decided to renew the effort, inviting Jake to consider entering into a conversation with his "aching gut."

Although his gut was not aching at the time of the session, Jake initiated a conversation with remarkable enthusiasm and uncharacteristic venom. Addressing his abdomen as "You miserable bastard, what the hell are you creating a fuss for? ... You are so pathetic ... shape up, goddamn it, relax and leave me alone!" Jake ended this outburst by saying to the therapist, "Wow, that felt good!"

The therapist then gently reminded Jake that conversations necessarily include more than one voice, inviting Jake to speak in the voice of his abdomen; "What does your aching gut want to say to you? ... Please speak to yourself in the voice of your abdomen."

Jake fidgeted, clearly irritated by this aspect of the process, but then slowly began to speak in a whining tone. "I can't go on like this..." The words faltered, Jake clutched his stomach, rolled into a fetal position on the couch, and started bawling and convulsing with tears. As this catharsis eventually subsided, Jake rolled back into his usual supine position and started laughing at himself, saying "I've just had the ugliest fantasy ... I'd like to take a knife and stab my boss in his tight young gut ... Then I'd like to roll him over and fuck him from behind!"

Again, for the purposes of our illustration, we do not need to describe Jake's odyssey much beyond this moment. We do need to note how subsequent sessions unleashed a range of aggressive feelings (including some aggression infused with sexual fantasies, and some reawakening of his sexual power occasionally inflected with hostile fantasies). Although the following formulation is inevitably simplistic, it seems warranted to suggest that Jake's anger was, so to speak, locked away in his abdomen and that the abdominal message of achiness was being disregarded or repressed. Jake was uncharacteristically explicit in his hostility toward his gut. When asked to speak to it, he essentially kept telling his abdomen to shut up. When Jake's abdomen was finally listened to, as a voice with its own message and its own prerogatives, only then did Jake reconnect with the full force of his sadness and his anger at the way in which his employment situation had disempowered him, and at the way in which this replicated previous experiences of disempowerment. Over the subsequent months, as Jake played and worked in this therapeutic manner, his sexual life with his partner flowered anew, his entire demeanor became more lively, and he successfully negotiated with his supervisor to be given a more challenging and interesting workload, at the same time developing a much more collegial and friendly relationship with this man.

• *Vignette C:* Layla came to therapy in her late thirties, presenting herself with a variety of problems relating to her role as a single parent, her anxieties about the future, and her recurrent difficulties with partnerships that did not prove enduring or satisfactory. Having only a high-school education, she had successfully worked her way up to a position of substantial responsibility in a moderately sized corporation. She had a pleasant complexion, but dressed conservatively as if to hide her attractiveness. She spoke of herself as the "ugly duckling." Her medium height and full figure were somewhat offset by rounded shoulders and a tendency to present herself with a timid and nervous demeanor.

In therapy, Layla spoke rather unemotionally about her young daughter, her worries about finances and future employment opportunities, and her occasional sexual liaisons with other women and more rarely with men. Her childhood experiences, as the eldest child in a large family without a father, were consistently glossed as "happy and unremarkable." Layla's therapist found it difficult to feel emotionally engaged with her. The various stories that repeatedly occupied every session might have elicited her empathy, but were narrated in such a flat manner that the therapist often found her mind wandering. Yet one feature of Layla's presentation caught the therapist's attention. As Layla spoke her shoulders would gradually, almost imperceptibly, stiffen and move upwards, while her head would droop slightly and her jaw lower toward her chest. This sequence would persist over several minutes, and then Layla would take a slightly deeper breath, and allow her shoulders to relax and her head to lift. Then, slowly the stiffening and lifting movement would begin again. This cycle had no discernible connection to the content of her spoken narrative.

Eventually the therapist asked Layla if she noticed her body moving in this manner as she spoke. Layla had no awareness of this and, even after several attempts by the therapist to draw her attention to her body's signal, the symptom remained entirely out of her awareness. Having observed some of the synergy methods used by Ilana Rubenfeld (Knaster, 1996; Rubenfeld, 2001; Rubenfeld & Borysenko, 2001), the therapist decided to conduct her own experiment. She invited Layla's participation, to which Layla readily, if somewhat compliantly, agreed. The therapist would sit quietly behind Layla as she talked about whatever was on her mind, and Layla gave the therapist permission to touch her whenever it seemed useful to do so.

After a period of getting accustomed to this somewhat unusual seating arrangement, Layla proceeded to talk in what seemed to be her usual manner, recounting episodes from her life with her daughter, her employment, and her occasional social engagements, in a repetitive and somewhat unemotional manner. As her shoulders started to lift, the therapist would gently and silently touch them, not leaving her hands on Layla's body for more than a second or so.

Initially, Layla seemed merely distracted by this procedure. She would cease her narrative, declare that she had "gone blank," and then return to whatever story she had been relating. Over a period of some weeks, however, Layla began to become aware of fleeting images arising as her shoulders were touched, and at the same time her sleep began to be disturbed by nightmares. As the duration and intensity of the images increased over time, Layla began to recall an aspect of her childhood that had been buried in amnesia. Her mother, whom she seemed to adore, would frequently cuff her about the head, without warning or any apparent reason. We do not need to document in any detail the subsequent progress that occurred over many months, except to indicate that a very different picture of Layla's childhood surfaced — a childhood in which she lived demeaned and in fear — and that her daily posture seemed to become more erect and her entire demeanor seemed to blossom. She completed a bachelor's and master's degree online, and is now director of a nonprofit agency for children who are victims of domestic violence.

• *Vignette D:* Jayden presented himself for therapy with what might be described as a "military bearing." His chest was large and prominent, his stomach tight and almost flat, and he walked with his pelvis tucked back and with a strut that involved minimal movement of his hips. When asked to take a deep breath, Jayden would lift his thoracic cavity up expansively, sucking in his abdomen, and raising his shoulders toward his ears. He had indeed been a ranking officer in the Marines, had attended a prestigious military academy, and had been on active operations. Now retired and in his forties, he was married and coming to therapy at the request of his wife, to whom he was devoted in a way that was almost obsequious. She had been in a very successful psychoanalysis which she had entered because she was upset over her infertility. She was now quite attuned to emotional nuances in her relationships and asked Jayden to seek therapy because she felt that "although he is a wonderful husband and spoils me delightfully, he has difficulty expressing his true feelings and being really intimate with me." She also complained that during their lovemaking he had a tendency to "just bang away at" her.

In therapy, Jayden did indeed seem somewhat emotionally nonreactive and inarticulate, claiming he did not understand "all these feeling things that my wife is so good at" and insisting that his life could be adequately governed by the "power of positive thinking." But he wanted to please his wife, wanted to become what she needed him to be. When describing their sexual intercourse, he was puzzled about her wishes and it became clear that, although he reached a climax with ejaculation, Jayden's orgasmic capacity was very restricted, and he was not experiencing full-bodied orgasming (cf., Chia & Arava, 1996; Ramsdale & Dorfman, 1985; Rosenberg, 1973).

The therapist sensed a tender side beneath Jayden's character defenses. For example, within a few sessions, Jayden confided something that he had told few other people. In a combat situation, there had been a miscommunication between him and the troops in the field, leading him to order an attack. It was later found that there was no enemy at the scene of the attack, but several children had been accidentally killed as a result of his command. Jayden's eyes misted over and watered as he recounted this episode, and it appeared that his entire body was taut, as if to hold himself against the onrush of his own feelings.

Subsequent to this tragic confession, the therapist invited Jayden to work therapeutically in a manner derived from bioenergetics and related methods (Boadella, 1986; Keleman, 1975a, 1975b; Lowen, 1976). For the purposes of our illustration, we do not need to detail the breathing exercises that were involved, accompanied occasionally by deliberate interventions on Jayden's posture and movement. Rather, what is important to note is that, over the months of therapy, Jayden softened in every aspect of his being. He became more emotionally attuned and articulate; his characteristic pattern of breathing deepened into his abdomen; his posture was visibly more relaxed; his orgasms became longer and more full-bodied, and his lovemaking became more playful and tenderly affectionate. It is pointless to speculate whether such changes in Jayden's personality might have occurred if he had been engaged in different therapeutic methods — such as those that are more committed solely to a "talking cure." Rather, the point is that characterological change — which is invariably a complex and gradual process — involves the somatic expression of a person's internal conflicts as much as it can be described in terms of mental representations.

Numerous cases such as these could be collected from the records of bodymind therapists. It could be argued that Lara and Layla both experienced a retrieval of repressed memories which had been somatically encoded, whereas this feature is less evident in the vignettes of Jake and Jayden. This, however, is merely a coincidence of the material selected for discussion — all these phenomena could be replicated in any gender, age, or ethnic group. These phenomena bear on the central issue of healing — namely, that it is not meaningful to view individuals in a dualistic manner that perpetuates the alienation of body and mind. Indeed, it may be suggested here that an approach to treatment that focuses solely on the one or the other aspect is doomed to be limited in its potential to heal, and this suggestion will be discussed further in the next chapter.

4
Healing Matters

We now need to offer some preliminary notes on the nature of healing, since this is a fundamental issue in which bodymind therapy offers a quite distinctive approach to the nature of benign change. The four vignettes described in the previous chapter illustrate three essential features of bodymind therapy. These three essentials are deeply interlinked (in actuality, they are aspects of one process). There are, of course, many valid ways to describe and consider them. For the moment, let us itemize them as follows:

- *Holistic discourse:* This is the importance of *honoring by listening* to all aspects of the individual and his or her ecology. In addition to listening to the mind's everyday systematic formulations, or even to the flow of free-associative movement within our reflective consciousness, there are processes of attending to, focusing on, and inviting into dialogue all the different "voices" that compose what European philosophers have called our "being-in-the-world" (the "beingness" of being human). In this way, new dimensions of awareness are developed. The impulses of our physicality — the living experience of our embodiment — can be treated as a "voice" and brought into this new awareness, just as much as the conceptual formulations that chatter in our heads or that come out of our mouths. No voice is assumed superior to another, and none is held in a relation of domination over the other.
- *Energy mobilization:* Although the notion of "energy" can embroil us in all sorts of difficult debates, we will use the term loosely here to refer to the shifts that are evoked by intentionally *breathing and moving* with awareness. For example, in the vignettes previously described, it may have been noted that moving or "breath-

ing into" whatever part of the body has called itself into conscious attention often seems to reconnect — so to speak — a body and mind that have become alienated or disconnected from each other. It is a practice whereby whatever meanings the body holds can be mobilized and thus brought into an expanded awareness of ourselves.

Appreciative connectivity: This is the importance of *touching with awareness.* In this sense, touching may occur as a physical palpation, an emotional engagement, or in some otherwise "energetic" sense. In this sense, healing never occurs without touch. Healing does not occur without a special process of connection, which is not that of dominative control or manipulation. Rather, it involves a distinctive dimension of wisdom. As we will discuss, healing requires processes of ethicality and an ontological momentum that is altogether distinct from the epistemological assumptions of the modern era, because it is not the effectualness of a change in the state of something (or someone) brought about by manipulation. Rather, healing is inherently a celebration of the liveliness of life itself.

Notions of healing

Influenced by modern science and the advances of allopathic medicine, we have become accustomed to notions of healing as being the result of a manipulative strategy that repairs or removes damaged or necrotic tissue, that involves adversarial action against invasive organisms and agents, or that merely relieves pain. As significant as these accomplishments may be, other philosophies of healing — whether complementary, alternative, Âyurvedic, osteopathic, chiropractic, homeopathic, naturopathic, or shamanic — have suggested the possibility of different perspectives (Albretch, Fitzpatrick & Scrimshaw, 2000; Ember, 2004; Saillant & Genest, 2007). It is not that bodymind therapy necessarily fits any of these treatment categories, but it does intimate a radically distinct approach to healing.

We might also note here that traditionally all healing has been understood as an act of God. This is explicit in the Torah, the Gospels and the Qu'ran. It is also evident in Dharmic, Taoic, and indigenous spiritual teachings. What is radically significant about this is that, in a slightly different terminology, these three essential features of bodymind therapy — holistic discourse, energy mobilization, and appreciative connectivity — are characteristics that can be ascribed to almost every nontheistic spiritual practice (Barratt, 2004a). There is a profound sense in which healing, as a process of being and becoming, is an inherently spiritual — even mystical — process. This will be discussed further in Chapter 16. Although we may not be able to specify quite what this entails, healing is a mobilization of the lifeforce and a manifestation in the present (some would say a "presencing") of our awareness of the lifeforce.

This healing potential of the lifeforce has been called the *vis medicatrix naturae*. It denotes the inherent potential of the bodymind to heal itself. This potential can be mobilized, but cannot be compelled (this is why, in a certain sense, it might be said that all healing is self-healing). Our awareness of it is a process of bodymind consciousness that is energetically or spiritually distinct from the subject/object, reflective self-consciousness that has traditionally been the exclusive notion of consciousness in the western world.

The distinctiveness of healing processes

It is for these reasons that healing is not a process of manipulation, not an instrumental act, a means directed toward an ends, or a goal-oriented procedure. It operates in the ethical and ontological context of interconnectedness, which is radically and dynamically distinct from the epistemology of subject/object (cf., Adorno, 1982). The latter is committed to dichotomies of mind/matter, method/outcome, agent/recipient or practitioner/client. The processes of healing elude these dichotomies. In this respect, it is easier to specify what healing is not, and then, in the remainder of this book, for us to point toward what healing might be. In a preliminary contemplation, three points might be noted here.

First, healing is not the avoidance of pain, nor even necessarily its palliation. The experiences of pain and the action of nociceptive mechanisms (the neural reception of injurious stimuli of which we are not conscious) are, after all, a signal both of the need for healing and of healing that is actively in process (Ornstein & Sobel, 1988; Waugh & Grant, 2001). Leszek Kolokowski (1989) provides us with a vigorous critique of the way in which the west has become a culture of analgesics; its ideologies failing to recognize the meaning of pain and instead, insisting on its immediate alleviation (cf., Morris, 1991; Rey, 1995; Sontag, 2001). Since pain is integral to life itself, and to the healing processes of life, it cannot be avoided. But it can be concealed, which is often the task pursued by modern medicine sometimes to the detriment of deeper healing processes.

Western cultures tend to value the concealment of suffering, rather than its confrontation. This applies not only to the physical level of bodily ailments, but also to the emotional level of intrapersonal and interpersonal ailments. It applies as well to the social level in which political programs are designed to hide poverty, injustice, and human degradation, rather than to eradicate their causes. The ideological forces of western culture push us toward what has been called the "narcotization" of life. This pertains not only to our use of substances, but also to our use of all the media of social commerce, entertainment and organized religion. It is an advocacy that curtails our potential for healing.

Second, healing is not the avoidance of death. Although it is not within the scope of this essay to explore the issue, it has been extensively shown that much of what is problematic about human egotism is founded on our determination to avert death (cf., Barratt, 2004a, 2004b). This is not only a Buddhist tenet. It has been discussed more recently in popular accounts (e.g., Becker, 1998; Kubler-Ross, 1997). And it has been given more intensive examination in an enormous range of works in the western philosophical tradition from Titus Lucretius Carus, to Karl Wilhelm Friedrich von Schlegel and Maurice Blanchot (cf., Crtichley, 2004; Derrida, 1996; Lingis, 1989). Western belief systems trenchantly define death as the opposite of life. This ideological failure to appreciate the inherent "deathfulness" of life itself leads us to conduct ourselves under the illusory possibility of a life without death, a life beyond death, and so forth.

To realize that deathfulness inheres to every moment in the liveliness of life itself is also to understand that our egotism establishes itself on an illusion that denies the interconnectedness of all things. Our egotism is founded on the illusion that it is itself autonomous, substantial and "really real" (Barratt, 2004a, 2004b). This realization leads us to the insight that genuine healing necessarily embraces death or, more precisely, genuine healing implies an awareness of the deathfulness of life itself. Healing requires the understanding of interconnectedness — the understanding that every moment of life itself entails death, that living is always a process of "deathfulness" and that destruction is inherent in every moment of creation.

Third, healing is not a procedure of political or sociocultural adaptation. It is not to be understood as a sub-genre of the various mechanisms of socialization and acculturation.

Although this mistake pertains conspicuously to the practices of psychiatry and the behavioral sciences, it is relevant to physical medicine also. Consider even the simple example of carpal tunnel syndrome. Suppose a doctor alleviates a typist's wrist pain caused by median neuro-pathy, only to enable him or her to return to a work-life that requires an entirely unnatural usage of the body. The work-life of someone who has to type for six, eight, or twelve hours a day, involves repetitive movements that are, sooner or later, bound to damage the individual's connective tissues in the wrists and elsewhere. To alleviate the pain of carpal tunnel syndrome certainly enables the patient to be "a pro-ductive citizen," earning a wage, and contributing to the profitability of the corporation. But in what sense is this really an act of healing? Its outcome primarily benefits the perpetuation of a social order that requires people to perform tasks that are entirely unnatural to their bodily constitution. It scarcely honors the body itself.

Getting "fit enough" to resume an unnatural task — or a task that is unnatural to the body in the manner or duration of its performance — is harmful to that body. A genuinely healing process would more plausibly involve a change in the social order such that every worker's daily routines could be varied in a manner that honors the versatility of each individual's embodiment. It is not an extraordinary feat of ima-gination to envision a social order in which no one had to perform the same damagingly repetitive task for the entire day. It is not difficult to imagine a society in which almost everyone did some of the typing that needed to be done, some of the ditch-digging, some of the plow-ing, some of the intellectual labor, and so forth. Such a society would be a more natural environment for the human bodies that constitute the labor force. But, of course, it would likely make less profit for the corporation's shareholders and the ruling-class.

The example of carpal tunnel syndrome could be multiplied many times over just in the realm of physical medicine. However, in the field of "behavioral disorders," the confusion between healing and adaptation is perhaps even more conspicuous. Sadly, the disciplines of psychiatry and behavioral science have often seemed to have a double-edge mission: either to equip individuals to fit within the social mainstream, com-placently fulfilling their role within the machinery of capitalist pro-duction, or to identify individuals who cannot fit within the dominant order and control, marginalize or eradicate their socially problematic behaviors.

Providing a critique of the collusion between the behavioral-psychiatric sciences and the dominant social order has been the important con-tribution of the so-called anti-psychiatry movement over the past forty years (Cooper, 1967, 1968, 1978). Throughout its history, psychiatry has too often been the discipline that controls and manages those who are socially or culturally marginal, dissident, different or disenfranchised (Szasz, 1984, 1989). Too often have the behavioral sciences provided pseudo-rationales for the use of force (physical and psychological) in controlling and limiting deviance from societal norms (Laing, 1960; Kleinman, 1988; Szasz, 2007a). From the use of asylums as punishment for disobedient women, to the use of electroshock and early drug treatments to oppress those who opposed the interests of the ruling-class, to the contemporary use of medications to enable individuals to conform better to everyday life in the corporations, in the dysfunctional family units, and in the military-industrial complex, the history is far from benign (Foucault, 1988). We will discuss this further in Chapter 15.

For now, let us simply conclude that adapting individuals to an oppressive social context is not equivalent to healing their ailments. Healing involves a process of personal and ecological growth, balance or harmony. Healing is not the machinations of a coercive socio-cultural order attempting to regulate its citizenry.

Freedom and presence

Although the notion may have somewhat old-fashioned nuances, it is worth reviving the term *ailment* in order to focus our understanding of the nature of healing processes (cf., Main, 1989). The ailment is not equivalent to a symptom or syndrome, nor is it equivalent to a "disease entity" or pathogenic phenomenon. It is also not the same as the patient's "presenting problem." Although such a presentation always points to the existence of an ailment, the patient is not always reflectively conscious of the ailment from which he or she is suffering — the ailment cannot necessarily be articulated in representational consciousness. An ailment is an affliction, a cause for complaining. In a sense, the ailment is an otherwise awareness of adversity — an awareness that is quite distinct from the formulation of a problem. The ailment speaks of adversity within the functioning of the bodymind; it is the voice of our bodymind expressing its suffering.

[As an aside, it will be noticed that the term "patient" is almost always used throughout this book, for much the same reason that the notion of "ailment" is used. The word comes via Old French from the present participle of the Latin verb *patī* — "suffer." It intrinsically connotes that the recipient of healing practices is in that circumstance because he or she is suffering. The derogatory medical connotation of being passively at the will of a physician's authority is a late development. By contrast, the term "client" means someone who is seeking services, and its connotation is entrepreneurial. The original status of a client is lowly and dependent. According to Ayto (2001), the term is derived from *cliēns*, which is an alteration of the earlier *cluēns* (the present participle of the Latin verb *cluēre*, meaning "follow" or "obey"). A client attends to the other's actions, and is unable to take independent action. By the seventeenth century, the term had broken free of its strictly legal usage and came simply to mean "customer."]

An ailment can be understood as a locus of stagnation in the natural healing processes of our embodiment. It represents a blockage in these processes — an interruption to the natural flow of healing. It is a fixation in the past or future that obstructs the flow of the present. Just as the growth of a tumor usually both indicates the past intake of toxins (or other invasive entities) and foretells the future death of surrounding tissues, there is a sense in which an ailment always presents as a matter that is oriented to past or future (or, more precisely, to the representation of past-futures). To describe this simplistically: Lara's tightness in her upper chest, neck and jaw, speaks the truth of her feelings about forgotten memories of fellatio with her brother; Jake's aching gut expresses unacknowledged anguish over his employment situation and anticipates the emergence of aggressive fantasies toward his boss; Layla's tensing shoulders express her childhood fears of unpredictable violence, which she unconsciously expects to recur in the future; and Jayden's entire posture and manner of movement conveys the trauma of his past, as well as his fears of deep tender feelings that might erupt into the future.

The ailment thus presents something that is, so to speak, a "blockage by repetition" that expresses the unexpressed. In our previous illustrations, the tightness, the aching, the tensing, the bracing rigidities all point to past pain and future fear that have not been processed or released into healing. These repetitions are the manifestations of avoidable suffering, grounded in unprocessed traumas from the past that instigate a fearfulness of the future. Healing is not a referential activity directed to redoing what was (as if that were possible), to indulging nostalgically or regretfully in what might have been, or to goal-oriented strategizing about what should be. Rather, it is the release of our being-in-the-world into the natural flow of what our being becomes when it is no longer blocked, stifled, constrained or curtailed. Healing is the mobilization of the lifeforce and a presencing of our awareness of this natural power.

Healing is thus an opening and a freeing. It is a process that releases us from the past-futures of what we think we know. This is why the distinction between representational knowledge (that re-presents the past and allows us to believe we can make predictions of the future) and the wisdom of awareness (as the presencing of our being-in-the-world to its processes of becoming) is so often invoked when trying to articulate the divergence between the compulsion of manipulated change and the freedom of healing processes.

Healing invites the awareness of presence, and health requires freedom (cf., Barratt, 2005). Contrary to what is commonly believed in western cultures (including those militaristic societies that tout themselves across the planet as the "defenders of freedom"), freedom and healing into health cannot be attained by coercion, cajolement or compulsion. Healing processes address the ailment, inviting its meanings to shift their mode of expression, inviting blockages to dissipate into the natural flow of our beingness. In this way, healing transmutes the adversity intimated by the ailment into alignment or attunement with the natural flow of spiritual-psychic-somatic energy. This is why insightfully wise practitioners have always taught that healing is the action of *ahimsā* — the presencing of Love.

5
The State of Emergence

We live in a period in which humanity is in the process of remembering or rediscovering itself as an aspect of nature. For the westernized technocratic world this is indeed an arduous, challenging and exhilarating process of rediscovery. Concomitantly, we live in a period in which we are rediscovering the divine, not as obedience to a judgmental and paternalistic God, but as the living Spirit that flows within each of us and all around us. Whether, or to what extent, these rediscoveries will avert the disastrous course on which the industrial and militaristic structures of transnational capitalism have embarked us remains to be seen. It is in this period that somatic psychology emerges, and there are three ways to frame the history of this emergence:

- First, there is a sense in which somatic psychology and bodymind therapy have been practiced for millennia. For they inhere to ancient spiritual traditions, and are evident in indigenous methods of healing. In this sense, the bodymind perspective was occluded by the industrial developments of western technology along with the cultural structures of globalized capitalism. Some commentators also argue that, even before those developments, this bodymind vision of the human being was ideologically opposed by the hegemony of specific forms of Christian and Islamic theology. In the twentieth century, the bodymind perspective was also systematically and ideologically obscured by the development of behaviorist, cognitivist and psychoanalytic psychologies (which tended to relegate the body to a secondary status, as was discussed in Chapter 1). But despite these adverse developments, somatic psychology and bodymind therapy are manifestations of ancient lineages of wisdom that are perhaps now, once again, coming into their own.

- Second, there is also a sense in which somatic psychology and bodymind therapy were initiated as an offshoot of western psychotherapy in the early twentieth century. This is the sense in which Wilhelm Reich (1897–1957), a brilliant and dissident student and colleague of Freud's, is sometimes honored as the "father" of somatic psychology. We will discuss this development further in Chapter 7.

- Third, there is a sense in which somatic psychology and bodymind therapy are very recent developments, emerging since the pivotal decade of the 1960s, beginning to blossom in Europe and North America very much in the last two decades of the twentieth century, and gathering strength as the twenty-first century gets underway.

Although it would be somewhat premature to write a history of the emergence of this discipline, this chapter will outline some of the main events that occurred through the twentieth century in terms of leading personalities and organizations. You will thus be offered a sketch of the current state of somatic psychology and bodymind therapy as a discipline. Although this sketch may seem a little dry, genealogical and bibliographic, it is important to outline the context in which somatic psychology moves toward its current significance. Then in Chapters 6 through 12, we will discuss seven of the most significant sources that have contributed to the emergence of this discipline.

The heritage of Wilhelm Reich

It is well known that Reich was one of several early psychoanalysts who diverged from what became Freudian orthodoxy (Sharaf, 1994). He became a student of psychoanalysis in 1919, his brilliance making a strong impression on Freud himself. He graduated to full membership in the Vienna Psychoanalytic Association a year later (Reich, 1967). Well regarded as a clinician and theorist, Reich published prodigiously. As a committed socialist, he focused intently on the socioeconomic causes of psychological suffering, and was a major advocate of sexual freedom and women's liberation (Reich, 1971a, 1971b). Early in his psychoanalytic career, he wrote extensively about the harm of sexual repression, about the damage done by the blocking of libidinal energy, about the way in which "character armoring" expresses and contains the patient's fear of orgasmically free-flowing energy, and about the so-called "vegetative" aspects of neurosis (Reich, 1980a, 1980b). He experimented with psychoanalytic procedures, having physical contact with his patients in order to work with their breathing, taking an active role in sessions by inviting them to modify their postural movements, and occasionally having them disrobe to facilitate visual access to the somatic manifestations of their ailments. Out of this clinical experience, he developed Freud's early ideas about libidinal energy into a theory of the lifeforce called "orgone," which has provided a theoretical foundation for some of his successors (Boadella, 1986; Reich, 1986). Reich's methods of therapy progressed through several phases as his career developed, and we will discuss these further in Chapter 7.

With the rise of Nazism (which he condemned as a political symptom of sexual repression), Reich fled first to Scandinavia and then, in 1939, to the United States (Martin, 2000). He had been expelled from the Communist Party in 1933 for his vehement critique of fascism (which later led him to call communism a form of "red fascism"). He was expelled from the International Psychoanalytic Association in 1934 because of his left-wing militancy and because of his therapeutic approach to the body (Reich, 1980a, 1980b). He had also provoked disfavor within the psychoanalytic establishment for his critical opinion of Freud's 1920 thesis on the *Todestrieb* or "death instinct." In the States, Reich continued to research his orgone theories as well as his distinctive style of bodymind therapy, which was called by several names depending on the phase of his career (Reich, 1961; Totten, 2003). He was the recipient of increasingly virulent attacks by orthodox psychiatrists, including some psychoanalysts. By 1956, he was jailed under the Food and Drug Act for making claims about the curative properties of orgone energy therapy. In Danbury Federal Prison, he was psychiatrically diagnosed as "paranoid with delusions of grandiosity, persecution, and ideas of reference." Reich was transferred to Lewisburg Federal Penitentiary where he died in 1957. In those years, and again in 1960, his books were burned by the United States' authorities, and it was not until late in the 1960s that further research on his work began to surface on both sides of the Atlantic.

Reich's legacy has been less conspicuously involved with orgone phenomena (as a treatment for diseases such as cancer and for other purposes), although there are some important developments in this direction. His legacy is more evident in the general development of bodymind therapy and especially of "bioenergetics" and related methods (Boadella, 1991; Totten, 2003).

In Europe, "vegetotherapy" and orgonomy, which are variants of Reich's pioneering work, developed more or less continuously in the postwar period and to the present. Ola Raknes (1887–1975), who had been psychoanalytically trained by both Karen Horney and Otto

Fenichel, became Reich's close friend and colleague in Scandinavia; he helped develop "character analytic vegetotherapy" and trained later generations of practitioners throughout Europe (Raknes, 2004). The significance of these Scandinavian developments in the history of somatic psychology cannot be underestimated. One of Raknes' students was Gerda Boyeson, who went on to develop her own influential style of "biodynamic" therapy (Boyeson, 1994; Boyeson & Boyeson, 1987). She, in turn, trained many practitioners who initiated their own styles of practice, one example of which is "organismic psychotherapy" developed by Malcolm and Katherine Brown. Another student of Raknes was David Boadella, who also trained in "psychosomatic centering" with Robert Moore in Denmark; in 1970 he established the seminal journal, *Energy and Character* (which is now published from Switzerland and Brazil), and he developed his own style of "biosynthesis" (Boadella, 1987). The influence of pioneers such as Raknes and Boyeson has provided a strong and almost continuous lineage of bodymind therapy practitioners in Europe.

However, Reich and his followers were not the only influences on the early development of bodymind therapy on the continent. From the late nineteenth century until World War II, Europe had intermittently developed a gymnosophist movement, including a controversial and marginalized ideology of naturist *freikörperkultur*. So in this sense the social context was prepared for a return to the body (cf., Grisko, 1999). We must note that there are other pioneers, who are sometimes overshadowed by Reich's notoriety, his extraordinary originality, and his prolific publication. For example, Elsa Gindler (1885–1961) and Heinrich Jacoby (1889–1964) met in Berlin in 1924 and proceeded to develop their own style of bodywork based on somatic sensitivity, experimentation with breathing patterns, and the cultivation of awareness. It is almost certain that Reich was influenced by their innovations, especially since his first wife was one of their pupils. However, Reich's major opus, *Character Analysis*, which began as a paper written in 1928 and first published as a book in 1933, does not credit their work directly (cf., Totten, 1998). Gindler and Jacoby influenced a generation of bodymind therapists on both sides of the Atlantic, including but not limited to: Charlotte Selver (whose work with sensory awareness has had a widespread impact, as we will soon discuss), Carola Speads (who developed a system of "physical re-education"), Ilse Middendorf (who developed her own style of "experience of breath" and founded training institutes in Europe and North America); as well as Lily Ehrenfried and Elaine Summers (whose methods of "kinetic awareness" became very influential from the 1960s onwards). It is also likely that Moshe Feldenkrais (who developed the "Feldenkrais Method") and Marion Rosen (who developed the "Rosen Method") felt the impact of Gindler's work, directly or indirectly; although the influence of Frederick Matthias Alexander (who developed the "Alexander Technique") on Feldenkrais is more evident.

Reich's heritage in North America has involved several diverse but interrelated developments. It seems that Reich asked Elsworth Baker to take responsibility for developing the practices of orgonomy, as the natural science of orgone energy and its functions. In 1967, the *Journal of Orgonomy* was founded, and the next year the "American College of Orgonomy" was established (cf., Baker, 1967; Baker & Reich, 1955; Reich, 1979). The interesting work done by this organization has limited its potential influence by its decision to train only medical doctors. It seems that the result of this policy has been the organization's declining influence on the field of psychotherapy. Parenthetically it may be noted that the same error was committed by the American Psychoanalytic Association, which diminished its impact by its early efforts to restrict training in the discipline to allopathic medical doctors; this was a policy which Freud himself opposed quite adamantly (Freud, 1926a).

The major standard-bearers of Reichian, post-Reichian and neo-Reichian, approaches to therapy in North America have been Alexander Lowen, John Pierrakos, Stanley Keleman, and Charles Kelley. Perhaps, more controversially, Fritz Perls might be included in this list since, although not considered a Reichian, he is believed to have had some training with Reich himself. Each of these developed their own approaches to therapy.

Lowen studied with Reich in the 1940s and 1950s, and went on to develop bioenergetics (sometimes known as bioenergetic analysis) with his then-colleague John Pierrakos. Lowen founded a major training institute in New York; over five decades, he published many popular and professional books, which continue to influence the field (e.g., Lowen, 1965, 1976, 1990, 1994, 2003, 2005a, 2005b).

Pierrakos initially trained with Reich and collaborated with Lowen. He later developed his own style of "core energetics" (Pierrakos, 2005). Totten (2003) suggests that development depended significantly on Pierrakos' ability to perceive energetic auras — although the ability is not required for training in this method. The development of core energetics was also connected with Eva Pierrakos' work with the spiritual practices known as "pathwork" (e.g., Pierrakos, 1993).

Keleman, trained with Lowen and has, since the late 1970s, become a major influence on the development of bodymind therapy internationally. He made an intensive study of emotional anatomy and developed his own system of "somatic-emotional therapy" and "formative psychology" publishing extensively on this topic (e.g., Keleman, 1975a, 1975b, 1986, 1987a, 1987b, 1999).

Kelley was Reich's student in the 1950s and proceeded to develop his own practice called "Radix," which is Latin for "root" and is his term for the energies of the lifeforce. Kelley established an institute and a training program dedicated to his approach, which is described in several publications (Boadella, 1991; Glenn & Müller-Schwefe, 1999; Kelley, 1974).

Perls is well known as a major figure in the development of "third force" psychologies, and his contributions will be mentioned further in Chapter 7. His career as a psychiatrist started on the fringes of the psychoanalytic mainstream in Berlin and South Africa. His first book, *Ego, Hunger and Aggression*, which was published in the 1940s, reflected this comparatively orthodox position (Perls, 1969). Coming to the United States in 1946, he trained briefly with Karen Horney (who has been called the "gentle rebel of psychoanalysis"). Later, he allegedly undertook some training with Reich, although the evidence on this point is equivocal.

Although commentators often refer to Perls as if he were the sole innovator of Gestalt therapy, it is clear that Laura Perls was a major but often under-acknowledged contributor. Not only was she a brilliant psychologist and psychotherapist with strong interests in Freud's early work, she co-founded the New York Institute for Gestalt Therapy, and continued to run it when her husband moved to the Esalen Institute in California. It is also well known that she co-wrote *Ego, Hunger and Aggression*, but was not credited as author. Her husband's seminal book, *Gestalt Therapy*, was based on his notes, refined by the psychological expertise of Ralph Hefferline, and extensively organized and written by the brilliance of Paul Goodman (who was much influenced by the here-and-now psychoanalytic methods of Otto Rank). Although Perls is undeniably a conspicuous figure in Gestalt therapy, the roots of this movement may be more appropriately traced to European phenomenology and existential psychology. However these developments are understood, it is clear that Gestalt therapy training flourished through the 1960s and thereafter, based on the insights and labors of not one, but many, practitioners. In some people's minds, however, Perls remains the totemic head of this movement — perhaps more for his charisma and therapeutic showmanship than for his theoretical acumen.

Much bodymind therapy has developed from this lineage. For example, Claudio Naranjo, who became a seminal figure in the human potential movement, trained and worked with Perls (cf., Naranjo, 2006a, 2006b); Jack Lee Rosenberg and Marjorie Rand's development of "integrative body psychotherapy" was influenced by Perls' work (Rosenberg, Rand & Asay, 1987); and, although perhaps not exactly falling within the definition of bodymind therapy, the innovation of neurolinguistic programming by Richard Bandler and John Grinder was also influenced by Perls' practices. In a sense, all these developments owe, directly or indirectly, to the Reichian heritage.

Notes on the context of third-force psychologies

It is far from the case that the emergence of somatic psychology and bodymind therapy emanates solely from the twentieth century's Reichian heritage, for there are many other interrelated developments that have changed — and are continuing to change — the shape of psychology. These developments began notably in the late 1950s and 1960s, and have accelerated since the 1980s.

To a greater or lesser degree, the establishment of a "third-force" in psychology — that is, approaches to the discipline that dissent from both cognitive behaviorist and psychoanalytic traditions — heralded a new interest in the holistic interconnectedness of body and mind (Bugental, 1964). At least initially, the emergence of these new approaches — which dubbed themselves humanistic — was perhaps more pronounced in North America than in Europe (cf., Van Kaam, 1960). When transpersonal perspectives became more popular — and Carl Jung's insights into the archetypal aspects of the human experience became developed and refined by articulate thinkers such as Marie-Louise von Franz, James Hillman and Marion Woodman — some people began to speak of a "fourth force" in psychology (Grof, 2000; Tart, 1992; Scotton, Chinen & Battista, 1996; Walsh & Vaughan, 1993). This will be mentioned further in Chapter 11. And currently, the burgeoning of multicultural perspectives in psychology is being called a "fifth force" (cf., Fay, 1996; Mio, Barker-Hackett & Tumambing, 2008).

Humanistic psychology not only presented a holistic vision that emphasized personhood, it focused on the meaning-seeking and creative aspects of the individual's growth (Moss, 1999). Its eminent figures in the 1960s were person-centered psychologists such as Carl Rogers, Clark Moustakas, and many others. This momentum embraced ambivalently some of the work of such dissident psychodynamic thinkers as Otto Rank, Erich Fromm, Karen Horney, Harry Stack Sullivan, Bruno Bettelheim and Erik Erikson. It also drew loosely from existentialist perspectives as articulated by writers such as Medard Boss, Ronald Laing, and Rollo May (Schneider, 2007; Schneider, Pierson & Bugenthal, 2002).

Perhaps more saliently, humanistic psychology gradually became quite committed to the theme of what has been called "height psychology" — in contrast to the "depth psychologies" that focused on the influence of unconscious factors over our functioning. Height psychology was initially championed most notably by Viktor Frankl during his period of cooperation with the Nazis (prior to his internment in a concentration camp). Under the influence of Alfred Adler's psychology (Adler, 1998), he emphasized issues of willpower, responsibility, and "spiritually-guided" self-determination (Frankl, 1937). Whatever the merits of such an emphasis (for example, it has had an important influence on the twelve-step recovery movement), it comprises a problematic break with psycho-analysis. Although often characterized as a dissent from the deterministic reading of Freud's discoveries, it is more based on an opposition to Freud's recognition of the body as the locus of the lived experience on which all other aspects of our psychology are grounded. Height psycho-logy is animated by a need to deny the wisdom of the body and its sexuality, upholding instead an idealization of moralizing notions about willpower. After Frankl's experience in the concentration camps, he developed this perspective in the 1940s and thereafter as his theory of "logotherapy" (Frankl, 1984). In many respects, height psychology is a precursor to the work of such influential psychologists as Abraham Maslow, the self-help movement with its Americanized ideology of what Dale Carnegie called the "responsibility assumption" (which suggests that you are fully responsible for the conditions of your life), and the more recent rash of enthusiasm for "positive psychology" (e.g., Seligman, Linley & Joseph, 2004). It is at the extreme positions of height psycho-logy that one can clearly see the philosophical pitfalls of the humanistic tradition. A moralizing overemphasis on individual responsibility and self-determination, although very concordant with the American values of individualism, leads to a socially conservative ideology. This ideo-logy overlooks the natural and cultural ecology — as well as the bodily foundation — within which personhood is constituted and on the basis of which individuals conduct their everyday existence (cf., Abram, 1997; Brown & Toadvine, 2003; Griffin, 1996; Prilleltensky, 1992).

Whatever the pitfalls of humanistic psychology — such as its tendency both to advance rather abstract, ideologically inflected or moralizing values, and to neglect the grounding of human experience in its embodiment and its ecology — the momentum of this "third force" has set the stage for the emergence of somatic psychology and bodymind therapy. Specifically, the call to return to *experience* as the focus of psychological inquiry, with its roots in phenomenological, existential, and organismic perspectives (the latter being represented by the work of Kurt Goldstein, for example), is profoundly significant. The renewed openness to Jungian insights, to energetic notions, and to the realm of transpersonal phenomena, has also made possible the return to somatically-oriented theories and therapies.

As we have defined it, somatic psychology and bodymind therapy have little in common with disciplines that are merely *about* the body. So, while there is interest in the findings of these disciplines, somatic psychology is radically different from enterprises such as psycho-somatic medicine (in the frame of modern allopathic science), rehabilitation medicine, sports psychology, human factors engineering, and all the associated disciplines that address the body in an objectivistic manner and attempt to improve its performance. It is also different from the sort of "mind-body medicine" that documents the effects of mental attitude on physical health (Harrington, 2009).

However, the emergence of a psychology based on the lived experience of embodiment has been contextualized by the development of several other sub-disciplines in the course of the last decades of the twentieth century. Let us note the following:

- Ecopsychology, for example, began to be more widely discussed in the 1990s, mostly following the prophetic insights of Alan Watts, Theodor Roszak and others. It is, in many respects, the twin of somatic psychology (Bronfenbrenner, 2006; Fisher, 2002; Louv, 2008; Metzner, 1999; Plotkin, 2007; Plotkin & Berry, 2003; Roszak, 2001; Roszak, Gomes & Kanner, 1995; Sabini, 2002; Sevilla, 2006; Watts, 1991, 1999; Winter, 1996). Indeed, it can well be argued that ecopsychology is an elaborated aspect of somatic psychology — or *vice versa.*

- Energy psychology, which is an ancient craft, has also begun to blossom in the past twenty years (e.g., Eden & Feinstein, 1999; Diamond, 1985; Feinstein, Eden & Craig, 2005; Gallo, 2002, 2004; Gerber, 2000, 2001; Martin & Landrell, 2005; Mollon, 2005, 2008; Oschman, 2000; Peerbolte, 1975; Sabetti, 2007). There are many variants and brand names in the clinical side of this sub-discipline, some of which have clear relevance to the advance of bodymind therapies. These include vibrational medicine, as well as a variety of methods with titles such as Thought Field Therapy, Emotional Freedom Techniques, Matrix Energetics, Tapas Acupressure Technique, Theta Healing, "healing from the body level up," Emotrance, Quantum Touch, and Quantum Energetics (cf., Mollon, 2005, 2008). It might also include the now popular technique of EMDR (eye movement desensitization and reprocessing). Some of the recent literature under this rubric has limited its significance by compromising its standards of science and scholarship. In this sense, the popularization of "energy talk" may not always be to the long-term advantage of somatic psychology and bodymind therapy — as will be discussed further in later chapters. However, the broadening of our understanding of the many sources, types or aspects of energy has obvious and profoundly important connections with somatic psychology, and there are now organizations of scientists doing important work in this area.

• Some of the more scholarly efforts that have contributed to the context that has nurtured the emergence of this discipline include: Ken Wilber's prolific writings on what he calls "integral psychology" (e.g., Wilber, 2000); Mikhaly Csikszentmihalyi's work on flow (e.g., Csikszentmihalyi, 1999); the introduction of scholarly writings on spiritual disciplines and practices that are grounded in bodily experience (e.g., Almaas, 2000a; Campbell, 1985, 1991; Ray, 2002a, 2002b, 2008); and the extensive and very substantial research currently being undertaken on the psychology of altered states and meditation (e.g., Baruš, 2003; Davidson & Harrington, 2001; Kabat-Zinn, 2006; Welwood, 2002). We will discuss some of these developments further in later chapters.

The current state of somatic psychology

The specific notion of the need for a distinctive discipline called *somatic psychology* seems to have come into currency sometime in the late 1970s or early 1980s, generated by a diverse and sizeable group of theorists and practitioners who were influenced by the sort of innovative work in somatics being undertaken in California at the Novato Institute, at the Esalen Institute, and at similar educational across the country. On the east coast, bioenergetic practices flourished during these decades, and various initiatives in somatic psychology blossomed across the United States and in Europe.

Thomas Hanna is often credited with proposing the term "somatics" to denote a specific disciplinary approach. Working in California in the 1960s, he had been very active in advancing the notion of somatics to refer to educational processes designed to "reawaken" — by methods of "functional integration" — the mind's potential to control movement, flexibility and health. With Eleanor Criswell Hanna, he founded the "Novato Institute for Somatic Research and Training" in 1975, started the journal, *Somatics*, and authored a couple of books on this topic (Hanna, 1993, 2004). Directly or indirectly, the Novato Institute provided a focus for a seminally influential group of somatic practitioners, psychologists and philosophers, including such eminent figures as: Frederick Matthias Alexander, who innovated his own technique for overcoming reactive, habitual limitations in movement and thinking (Alexander & Barlow, 2001; McGowan, 1997); Alexander Lowen, whose bioenergetic work we have already mentioned (e.g., Lowen, 1976, 2003); Moshé Feldenkrais, who studied the work of Alexander, Gindler, Jacoby, as well as the Armenian mystic George Ivanovich Gurdjieff (cf., Cravioto, 2007), and later developed his own method of awareness through movement (Feldenkrais, 1981, 1991, 2002, 2005); Ida Pauline Rolf, who founded the structural integration method (known commonly as "Rolfing") of deep soft tissue manipulation to achieve postural release (Rolf, 1989, 1990); Ilana Rubenfeld, who developed her own "synergy" method of combining talk and touch in a healing process (Rubenfeld, 2001; Rubenfeld & Borysenko, 2001); and Charlotte Selver (Littlewood & Roche, 2004; Selver & Brooks, 2007). Each of these individuals has had a major impact on the expansion of wisdom within each of their particular domains, and this list is far from exhaustive. This then constituted a major impetus in the contemporary development of somatic psychology and bodymind therapy.

We might make a special note of the work of Selver, who taught at the Esalen Institute from 1963 onwards and, despite the fact that she published little, had a very extensive influence on the development of the "human potential movement" (cf., Littlewood & Roche, 2004). Selver believed in the importance of trusting organic processes, and was an ardent advocate of what she called "sensory awareness." She had trained with Gindler and Jacoby in Berlin in the 1920s, coming to the United States in 1938 and establishing the "Sensory Awareness Foundation" in 1971. The list of people whom she taught or influenced in other ways is large, and includes such eminent figures as Erich Fromm, Fritz Perls, the maverick Zen philosopher Alan Watts , Moshé Feldenkrais, Ida Rolf, Don Hanlon Johnson, Judyth Weaver, and many others. The impact of the sensory awareness movement on today's influential practitioners of somatic psychology, such as Ron Kurtz, Peter Levine, Pat Ogden, Susan Aposhyan, and Christine Caldwell, cannot be overestimated.

The prestigious Esalen Institute was founded in 1962 by Michael Murphy and Dick Price in order to support innovative and multidisciplinary education in areas often neglected or underemphasized by traditional academic programs. It has provided, and continues to provide, an extraordinarily important forum in which these nascent initiatives in somatics, somatic psychology and bodymind therapy could be nurtured (Anderson, 1983). Not to belittle the many other important developments of these disciplines in Europe and elsewhere (nor to discount the importance of Esalen's east-coast counterpart, the Omega Institute, which was founded in 1977), it is suggested that the significance of the Esalen Institute in the historical emergence of somatic psychology and bodymind therapy can scarcely be over-estimated (Kripal, 2007; Kripal & Shuck, 2005). It has empowered these disciplines to grow steadily through the 1970s and into the present, and it provided a non-sectarian forum for this growth.

Alongside this flowering of developments outside the academy, somatic psychology began to be introduced in some of the more visionary institutions of graduate education. In 1983, Don Hanlon Johnson, a leading figure in the field, established the first master's degree program in Somatic Psychology at Antioch University, transferring it a year or so later to the California Institute of Integral Studies (Johnson, 1993, 1994, 1995, 1997, 2006; Johnson & Grand, 1998). Johnson is unlike many of the more entrepreneurial leaders in this field, in that he never developed his own brand of practice; however, his influence on the professional development of somatic psychology in North America is enormous (e.g., Johnson, 1993, 1994, 1995, 1997, 2006; Johnson & Grand, 1998).

In North America, other master's degree programs in somatic psychology or in closely related subjects were gradually initiated at several other regionally accredited institutions. In 1984, Christine Caldwell, who has published valuable material in this field, established the master's degree program at what is now Naropa University, which now has a strong tradition of contemplative education as envisioned by Chögyam Trungpa (the Tibetan Buddhist tülku who established "Shambhala" training across Europe and the Americas). Accredited master's level training in somatic psychology is available at other graduate institutions including: the Institute for Transpersonal Psychology, with its embrace of a range of spiritual disciplines; the John F. Kennedy University which, like South-western College, promotes an emphasis on transformational education; and the Saybrook University, with its history of foundational work in existential and phenomenological psychology, as well as its connection to the somatic work of Thomas and Eleanor Hanna, and its eminent contemporary faculty such as Stanley Krippner (e.g., Krippner, 1992) and Amedeo Giorgi (e.g., Giorgi, 1970, 1985). There are also a number of other accredited institutes and universities — as well as unaccredited ones — that have established courses and programs in somatic psychology and bodymind therapy, or as a specialized track within a clinical degree program.

In Europe, the situation is somewhat parallel. Following the lineages of Reich and his associates, as well as of Gindler and Jacoby with their many students, many new initiatives surfaced across the continent. Additionally, there were innovations that seem to have little direct connection with these lineages. One example of this would be the development of what has been called "Eutony" by Gerda Alexander, who moved from Germany to Denmark in 1929 and spent her career experimenting with methods of movement and awareness to enhance personal ease, freedom, and wellbeing (Alexander, 1981). Today, across the field of somatic psychology and bodymind therapy, there are many European training institutes that operate outside of academia, and a small number of universities that have established training programs in this field.

If we summarize what we have surveyed thus far, there is a large number of initiatives and training programs that fall within the rubric of somatic psychology and bodymind therapy. One can easily review the website of the European Association for Body Psychotherapy (see www.EABP.org) or of the United States Association for Body Psychotherapy (see www.USABP.org) in order to get a sense of the scope of what is merely a segment of the practitioners and training organizations in this field.

As has been mentioned previously, these include all the various Reichian, neo-Reichian and post-Reichian offshoots such as: character analytic vegetotherapy, orgonomy, biodynamics, organismic psychotherapy, psychosomatic centering, bioenergetics, core energetics, biosynthesis, somatic-emotional therapy, formative psychotherapy, "Radix," certain aspects of gestalt therapy along with integrative body psychotherapy, and others. Perhaps missing from this list are several other European initiatives including but not limited to: Lillemor Johnsen's respiratory therapy; Lisbeth Marcher and Erik Jarlnaes' work on bodynamics (Macnaughton, 2004); the development of therapeutic eurythmy as an expressive movement art originating in the anthroposophical work of Rudolph Steiner and developed by Marie von Sivers (Steiner & Usher, 2007); Luciano Rispoli's school of functional psychotherapy (Heller, M., 2001; Rispoli, 1993); and Jay Stattman's unitive psychotherapy (Stattman, 1989).

There are also many other offshoots of the sensory awareness lineage that developed out of Gindler and Jacoby's pioneering efforts, including:

somatic sensitivity (later somatic awareness), Feldenkrais method, Rosen method, physical re-education, and kinetic awareness. Perhaps missing from this list are several other North American initiatives including but not limited to: voice and body dialogue methods (cf., Griffith & Griffith, 1994; Rous, 2006; Stone & Stone, 1998); "Authentic Movement" initiated by Mary Whitehouse as a "movement-in-depth" method (Adler, 2002; Chodorow, 1991; Pallaro, 1999, 2007); Christine Caldwell's influential "Moving Cycle" methodology for which training is also offered in Europe (Caldwell, 1996, 1997; Lewis, 2002); the Pesso-Boyden system of psychomotor therapy (Pesso, 1969, 1990); Emilie Conrad's "Continuum" method of conscious and creative movement practices (Conrad, 2007; Gintis, 2007); Bonnie Bainbridge Cohen's methods of bodymind centering (Cohen, 2008); and other integrative therapies based on a specific bodywork method such as "Lomilomi" (Calvert, 2002; Chai, 2005, 2007). The influence of mindfulness meditation practices on the development of this movement has also been considerable (e.g., Kabat-Zinn, 1990, 2007). Dance therapy and movement therapies have also been influential in this respect, and various contemporarily eminent figures such as Christine Caldwell, Susan Aposhyan, Marjorie Rand, and Judyth Weaver, trained in these disciplines.

There are, indeed, almost innumerable hybrids and variants in the field of bodymind therapy. There is also an egregious tendency for entrepreneurially minded practitioners to combine methods, thus creating a minor variation on whatever they learned elsewhere, and then to market this as their own "new" and completely different brand. This makes for much confusion, and it becomes hard for the prospective patient, client or student, to be well informed about the modalities in which they might be interested.

However, five somewhat distinctive schools of bodymind therapy must be specifically mentioned here, since thus far they have not been discussed in any detail, and because they perhaps constitute some of the most influential approaches to the field, not only in North America but worldwide.

- Eugene Gendlin, who had worked with Carl Rogers in the 1960s, developed the method of "focusing" and "focusing-oriented psychotherapy" (Gendlin, 1982, 1997, 1998). Perhaps more than any other individual, Gendlin has provided a theoretical base for bodymind therapy in the philosophical tradition of phenomenology.
- Peter Levine originated the therapeutic methods of "somatic experiencing," which are based on psychophysiological trauma theory

concerning the immobility response, which is triggered in prey animals when danger is perceived (Levine, 2008; Levine & Frederick, 1997). Training in these methods is now available throughout the Americas, in Europe and parts of the Middle East. There are several other therapies closely related to somatic experiencing; one example of this is Babette Rothschild's "somatic trauma therapy" (Rothschild, 2000, 2003).

- Ron Kurtz originated the Hakomi approach to therapy, which draws systematically on a number of sources including bioenergetic and gestalt methods, sensory awareness, structural and psychomotor bodywork, focusing and neurolinguistic programming (Kurtz, 2007; Johanson & Kurtz, 1991; Kurtz & Prestera, 1984). Hakomi is a Hopi word, used by Kurtz to designate the healing principles of mindfulness, unity, holism, *ahimsā* or nonviolence, and organicity. Training in Hakomi is now available throughout the Americas, in Europe and some parts of Asia and Australasia. There were others involved in the formulation of Hakomi who later developed their own systems; one example of this is Pat Ogden's "sensorimotor psychotherapy" (Ogden, Minton & Pain, 2006).

- Arnold Mindell, a Jungian analyst, originated "process-oriented psychology" (which is often known as "process work"), which is based not only on his transpersonal studies but also his interest in quantum physics, Taoic spirituality, and subtle conditions of consciousness. A prolific writer and educator, Mindell's work has provided the framework for training institutes worldwide (Mindell, 1982, 1985a, 1985b, 1987, 1988, 1991a, 1991b, 1992, 1993, 2000, 2007).

- Finally, a variety of primal therapies need to be mentioned. In relation to the practice of bodymind therapy, perhaps the most important of these is the "holotropic breathwork" developed by Stanislav Grof, after he had relinquished his pioneering efforts with psychedelic psychotherapy (e.g., Grof, 1988, 1998, 2000, 2007; Grof & Bennet, 1993). This primal category includes modalities such as "primal scream," primal integration and rebirthing. Its theoretical base has obvious connections with the issues of birth trauma and the whole topic of prenatal and perinatal psychology, about which there are ongoing controversies (McCarty, 2008).

Today, many, but not all, of the themes and threads of bodymind therapy are affiliated with either the European Association for Body Psychotherapy (EABP) or the United States Association for Body Psychotherapy (USABP). The EABP was founded in 1989, following the first European Congress for Body Psychotherapy, which was organized by Jacob Stattman, Malcolm Brown, Bjorn Blumenthal, and Don Hanlon

Johnson. Approaching the end of the first decade of this century, it currently boasts around seven hundred professionally accredited members, over fifty organizational members, and about two dozen affiliated training institutes (including members from five countries outside of Europe). Amongst many other activities, it maintains a credentialing function, offers a range of opportunities for training and professional education, promulgates a fine code of ethics, and publishes a regularly updated bibliography of literature in the field of body psychotherapy. The USABP, founded in 1996, is a smaller and in some ways weaker organization that is currently growing, and that already has its own conferences biennially and its own professional journal.

Finally, in considering the state of emergence of somatic psychology and bodymind therapy, it must be mentioned that there is a large, and rapidly expanding body of literature in this field — a significant portion of which will be referenced in this book. Students, and professionals without any familiarity with this field, often ask: *Where to start?* I typically recommend three or four recent books:

- Susan Aposhyan's *Body-Mind Psychotherapy: Principles, Techniques and Practical Applications* (2004). Although somewhat slanted toward her own version of practice, this is a wonderful book for undergraduate and graduate students wishing to get acquainted with the field, as well as for professionals who are unfamiliar with this field and wish to understand all that it might offer their theoretical thinking and their clinical practice. One of the virtues of this book is that it has many clinical illustrations, and is relatively strong on the connection between bodymind therapy and recent advances in attachment theory and in the neurosciences.
- Edward Smith's *The Body in Psychotherapy* (2003) is an excellent companion to Aposhyan's work with a strong clinical orientation to the organismic perspective.
- Nick Totten's *Body Psychotherapy: An Introduction* (2003) is also an outstanding text for beginners. Totten has a strong understanding of European developments, which balances Aposhyan's emphasis (since she lives and practices in the North American context). Totten's subsequent anthology, *New Dimensions in Body Psychotherapy* (Totten, 2005) is also very helpful for the professional wanting to understand something about the range of approaches in this field.

After these volumes, students need to pursue whatever more specialized avenues of inquiry in which they might be interested. Hopefully, the extensive references in this book will provide a guide to this endeavor — indeed, one intention of this work is that it will provide a guide for any reader who wishes to undertake further studies in this field. The larger bibliographies provided by the EABP and the USABP are — of course — imperfect (in that they omit some important material) but are nonetheless very helpful.

Section II

Sources: Ancient and Contemporary

Critics might suggest that it is premature to designate a "new discipline" called *somatic psychology* (with bodymind therapy being its applied aspect). Linda Hartley's useful text by that name seems to be the only other English language book currently in print with this term in its title, and — perhaps wisely — she avoids confronting this criticism. Instead, she offers us a lively exploration of the interface of psychotherapy and psychological theory with the somatic practices of bodywork and movement therapy (Hartley, 2004). Yet the discipline of somatic psychology does exist, at least by name. Indeed, it could be argued that it has existed since ancient times — although without the name — and that it was temporarily eclipsed by the twentieth century's development of psychologies that ignore the experiential voice of our embodiment, relegating the body to a mere object that the mind attempts to control.

The question that remains — the question that this volume addresses — concerns the present status of this discipline, and its future potential within the spectrum of human knowledge and the panoply of healing practices.

As we survey the field as it now stands, it is evident that somatic psychology is not yet a cohesive or well-integrated discipline. From a formal standpoint, somatic psychology is still inchoate. So it cannot be claimed that this endeavor has yet achieved the honorific status of a well-integrated synthesis of previous developments. What is currently being achieved, and what I believe holds great potential for the future, is a syncretic momentum that is now undermining the modern era's ways of thinking about the human condition.

syncretic development is defined as one that brings together diverse themes and threads to blend them into the warp and woof of new fabric.

It draws creatively from many sources, but has not yet woven them into an internally coherent or seamless system. This then is the situation in which we find somatic psychology; several tributaries pointing toward a single initiative that has yet to achieve its potential for cohesive momentum. Accordingly, in this section, there will be a brief review of each of what may be considered the seven sources by which, and from which, this distinctive new discipline is currently emerging.

In this second section Chapter 6 will discuss the foundation of somatic psychology in Freud's work, particularly focusing on the notion of libidinality as an energy theory. Chapter 7 will examine the extent to which the pioneers of psychoanalysis defined an approach to understanding the human condition that I am calling *somatic psychodynamics*. Chapter 8 will briefly survey a wide range of philosophical and cultural trends in the twentieth century in order to show how the importance of the experiential body has been both forgotten and rediscovered. Chapter 9 will show how the western world has traditions of bodywork that somatic psychology and bodymind therapy can build upon. Chapter 10 investigates the way in which a wide variety of Asian disciplines have come to influence the philosophy and culture of the western world, and how these disciplines are changing our understanding of the human condition. Chapter 11 looks specifically at the highly significant yet still controversial practices of shamanic and transpersonal psychologies. And Chapter 12 provides a sketch of some of the recent advances in the neurosciences that validate the perspectives of somatic psychology and that thereby contribute powerfully to the mandate of bodymind practices.

Psychoanalytic Discoveries

The popular and professional image that most people have of Sigmund Freud — at least at this point in history — is definitely not that of a somatic psychologist and far from that of a bodymind therapist. But this assessment neglects both the extent to which Freud's methods subvert the epistemic assumptions of the modern era, foreshadowing radically new ways of thinking about the human condition, and the extent to which his version of psychoanalysis is, in certain respects, the prototype of a somatic psychology. The contemporary failure to appreciate Freud's psychology in this manner is based on three factors.

First, our educational institutions typically utilize only secondary texts for an understanding of Freud's psychology. Such works invariably depict psychoanalysis in terms of its three main interlocking or overlapping theories of psychological functioning: the so-called topographic model (of consciousness, preconscious, and unconscious "regions" of the mental functioning); the structural-functional model (in which the organized or ego aspect of the mind strives to construct workable compromises between id impulses, superego prohibitions, and reality as it construes it); and the various theories of self and so-called "object-relational" representations.

Each of these models, in different ways, systematically displaces the focus of psychological inquiry away from the lived experience of embodiment. In the topographic model, the role of the body is rendered in a complex theory of "thing-presentations" (versus "word-presentations"). Moreover, the spatialized depiction of the mind (as having conscious, preconscious, and unconscious regions) necessarily obscures the dynamic and processive conditions of libidinality, which we can define as the erotic energetics of the living experience of our embodiment. In the structural-functional model, the body becomes a machine-like thing the ego has to deal with — as an asset or as a liability. In the self and object-relational model, the body is presented predominantly in terms of the mind's ability to represent it.

Although these statements are simplifications, and may also be over-generalizations, they convey the fact that secondary texts describing Freud's psychology almost invariably cast it in terms that are con-cordant with the discourse of the modern era. If the libido theory is addressed at all, it is almost always seen as merely a theory of our drive to commit certain sort of sex acts, or as a theory of develop-mental phases in the child's preoccupations. This reductive and ideo-logically distorted rendition of what Freud actually said again serves to make psychoanalysis concordant with a masterdiscourse from which it dissents.

Second, even if we dispense with secondary texts, the ubiquit-ous failure to appreciate the radical implications of Freud's work — both as heralding postmodern thinking and as establishing somatic psychology — is based on a biased appreciation of his approach to the human condition. If Freud's writings are not read in the original German, then most English-speaking students use Strachey's tran-slations (commonly known as the "Standard Edition"). These have been justly criticized for presenting Freud's thought in a static and reified manner. For example, the more phenomenological *Ich* meaning "I" is rendered as "ego" or as *the* ego — a very substantive and hypo-statized depiction. The shifting of our energies (in German, *Besetzen* meaning to occupy, engage, fill or invest) is translated, in pseudo-technical language, by the Greek term *cathexis* — again making it seem remote from the lived experience of our embodiment. Problems such as these have motivated several commentators to call for a rereading of Freud in the original German, or for an effort at retranslation (cf., Barratt, 1984; Bettelheim, 1983; Derrida, 1985, 1989, 1999).

This issue of translation bias is compounded by a problem of selectivity in that, perhaps with the exception of the 1900 *Die Traumdeutung* (The Interpretation of Dreams), students tend to focus almost exclusively on Freud's post-1914 writings. Indeed, the con-temporary psychoanalytic world is divided as to whether they favor the so-called "metapsychological papers," which were written between 1914 and 1918, *Beyond the Pleasure Principle*, which was written in 1920, or the two major structural-functional expositions of 1923 *The Ego and the Id*) and 1926b (*Inhibitions, Symptoms and Anxiety*) as the apogee of Freud's endeavors. The relative neglect of pre-1914 writings has serious consequences in that it overrides the almost anarchic implications of Freud's methods of inquiry as liberation psychology (this notion will be discussed further in Chapter 15). Freud's post-1914 writings have a tendency to be preoccupied with systematic theorizing, and in certain fundamental respects this tendency rescinds the vivacity and the radical edginess of his earlier work.

The third factor responsible for the failure to appreciate Freud's radicalism is that almost all the post-Freudian theorists spent the twentieth century trying to abrogate the epistemically disruptive dimensions of his work. In Chapter 1, we sketched how the five or so major strands of today's psychoanalytic orthodoxy all variously betray the radical potential of psychoanalysis (again: the structural-functional tradition; Kleinian psychoanalysis; the various object-relational, relational or interpersonalist endeavors; Kohutian self-psychology; and Lacanian psychoanalysis). This betrayal comprises a reworking of psychoanalytic tenets in conformity with the epistemologies of the modern era (Cartesian, Kantian, neo-Kantian, Heideggerian or Saussurean), and specifically a disregard for the bodily grounding of human experience, which Freud had announced in his writings on libidinality (and which is evident in his manner of clinical practice).

The understanding of Freud's notion of libidinality is crucial to an appreciation of his work as the formal beginning of somatic psychology. Even before he had completed his so-called metapsychological papers — with their ambition to systematize psychoanalytic theorizing — and long before he had penned *The Ego and the Id* — Freud announced that his newly discovered method of interrogating consciousness opens us to "a critical new direction in the world and in science" (Freud, 1916–1917, p. 15). This method of free-associative discourse is essentially a process that reinvigorates reflective consciousness in relation to the unconscious that it represses from itself. Freud was confident of the significance of his new discipline precisely because — as he later wrote in Max Marcuse's encyclopedia of sexology — he understood psychoanalysis to be foremost, and most significantly, this method for listening to the voice of the repressed unconscious dimension of our own being-in-the-world (Freud, 1923b). What I believe was clear to Freud prior to about 1914 (and what became buried beneath his later preoccupations with systematic theorizing) is that a method for listening to the voice of the repressed necessarily and foundationally entails a process of listening to the voice of our embodiment (Barratt, 1984, 1993).

The division of Freud's career into his pre-1914 and post-1914 writings is a useful heuristic (but, as with any heuristic device, it is no more than that). It is known that Freud's temperament became less optimistic during and after World War I. It is also known that his painful struggle with cancer worsened around this time. No doubt these experiences — along with his conflicts with several of those who had been prominent supporters in the early years of psychoanalysis — intensified his wish to establish psychoanalysis as what he would later call "a natural science like any other." This ambition entailed a systematization of theory, a restoration of the precepts of modern science, and a gradual break with any emphasis on the process of listening to the voice of our bodily wisdom. The latter implied that the notion of libidinality be retained more or less in name only, if at all.

From his earliest discoveries, Freud understood mental life to be composed not only of representations (of self, of the body, of other people and things, and so forth) and the transformations between them, but also of something that occurs otherwise than the functions of representationality. He wrote variously of *Intensitäten, Besetzungen, Triebe* — of intensities; of engaging, occupying, distributing, casting or investing; and of impetuses, urges, driving propensities, impulses, inclinations, likings and desires. In sum, he tried to write about the momentum of the lifeforce within us, and he was aware that the lifeforce itself can never be captured in or by our mental representations, our narratives and our egotistic ambitions. Although, as an allopathic physician, he initially tried to anchor these dynamic events to neuronal activities; yet, as we will now discuss, he came to write of libidinality as having a sort of *both-and* yet *neither-nor* status that breaks the bounds of traditional scientific thinking.

On the one side, it is necessary to understand our psyche as composed of something above, beneath, and beyond a system of representations and their transformations (which form the narratives of all that we reflectively believe ourselves to be). In this sense, libido is discussed as if it were pure energy emanating from the materiality of our being; it is not itself meaningful but is, so to speak, the "juice" or Spirit that propels representational meaningfulness.

On the other side, libido is discussed in a manner somewhat similar to a contrarian version of Franz Brentano's notion of intentionality (Jacquette, 2004), or even Heidegger's notion of "care" as *freige-bende Sorge* (Heidegger, 2008; Kleinberg-Levin, 2005). It bestows its own modality of meaningfulness; yet this is not a meaningfulness that can be fully and adequately designated within the structures of representation. In this sense, libidinality is not only the unpredictable and uncontrollable spontaneity of the lifeforce, it is also a mode of meaningfulness that struggles for expression, yet is ineluctably otherwise than the modality of mental representation. The libidinal lifeforce is always, so to speak, "in" but not "of" the representation of meaningfulness. Therefore, it is "betwixt and between" — a liminal notion. This liminality of a power that is both *both-and* (both material and immaterial) and yet *neither-nor* (neither meaningless nor a meaning that can be represented) is, of course, utterly incomprehensible — at least within the epistemic coordinates of the modern era — and the implications of this have been discussed in more detail elsewhere (Barratt, 1993, pp. 134–182).

As Efron (1985) and others have demonstrated, instead of wrestling with the challenging implications of this, later psychoanalysts adopted several strategies. The notion of libido would be dispensed with entirely, or retained only in name, which makes "psychoanalysis" entirely a psychology of representations and their transformations (the repetition or revision of stories that form all that we reflectively believe ourselves to be) accompanied by a therapeutic practice that keeps itself wholly "in the head." Alternatively, the notion of libido could refer to the biological endowment of the individual, in which case "libidinality" comes to refer merely to normative phases in the child's bodily preoccupations (oral, anal, phallic and genital), and "libido" comes to refer simply to the individual's interest in sexual activity. As important as sexual activity is, the latter conceptualization denies the holistic character of our embodiment. Libidinality, as Freud presents it, is not "just about sex" as we customarily think about sex. Rather, it concerns the erogeneity of our entire body, the sensual body. We can understand this to mean the sexual body only if we also understand that everything is sexual. Libidinality is this "bodily otherwise" that is the voicing of our embodied being-in-the-world. It is a kinesis of a flowing momentum, the brio of life — the *élan vital* — that is irretrievably different from the formulations of our representational mentality. It is surely noteworthy that Freud's writings on this topic are contemporaneous with Henri Bergson's 1907 essay on the lifeforce or élan vital (Bergson, 2009; Deleuze, 1990). Libidinality is, under another description, our spirituality incarnate — the foundational process of our desire. Akin to the Asian notions of *prāṇā* or *chi* (which will be discussed in Chapter 10), libidinality is our lived experience of embodiment (cf., Atreya, 1996; Frantzis, 2008; Gopi, 1997; Niranjanananda & Niranjanananda, 2002; Ramacharaka, 2006). It is life's longing for itself.

The radical potential of psychoanalysis is its capacity to deliver a method of interrogating the psyche in a way that is both scientific and emancipative (cf., Barratt, 1993). The revolutionary essence of Freud's discovery is the method of listening, called free-associative discourse, which opens the structures of representational thinking to the voicing of desires that cannot be formulated in representational thought; voices that are otherwise than the law and order of what is commonly called the mind. This is the discovery of the unconscious — the discovery of a way of listening to the voices of our being-in-the-world that have been repressed from, and are incomprehensible to, the narrative formulations of representation.

Late in his life (significantly after the formulation of the topographic model, the structural-functional model, and all the various theories of self and object-relations), Freud declared that the only aspects of his work that would have lasting value were two books written prior to 1914. This can be taken as a retrospective recognition that, above and beyond all else, his method of interrogating the representations of consciousness is at the kernel of his originality. It might also be indicative of Freud's awareness that it was this aspect of his discoveries which would be avoided by most of his contemporaries and by those schools of psychoanalysis that developed after his death.

Yet even while his revolutionary originality was being obscured by his own theoretical formulations and by those of his contemporaries, Freud formulated his 1933 (*Gesammelte Werke*, p. 86) slogan for (re)integrative healing: *Wo Es war, soll Ich werden*. The slogan was — in my opinion — poorly translated in the rather mechanistic phrasing "Where Id was, there Ego shall be" (*Standard Edition*, p. 80). It is far better translated as *Where it was, should I become*, which is not only closer to the German, but conveys the more dynamic, organic and aspirational aspect of the message. Ten years earlier, Freud had trenchantly declared that his discipline had amply proved that the human "I is foremost a bodily I" (1923a, p. 255). Even while the insight into human experience as grounded in our embodiment was being eclipsed by the industry of "Freudian formulations," Freud's commitment to somatic psychology as the way of the future is still discernible. Freud was, first and foremost, a somatic psychologist, and his contributions stand at the head of this nascent discipline.

Somatic Psychodynamics

Although beyond the scope of this book, it would be interesting to trace the history of the psychoanalytic movement as a series of "body phobic" reactions to Freud's original discoveries. It can certainly be demonstrated that the earliest reactions were essentially a repudiation of the significance of libidinality, and that such a fear-based conservatism continues today. This repudiation operates either by disputing the significance of the sexual body (and hence avoiding the notion of libidinality almost entirely), or by conceptualizing Freud's discoveries in terms of a theory of "sex acts" (and hence avoiding almost entirely the practice of listening to bodily experience). Here it will be argued that these two seemingly contrary positions are both actually indicative, in quite different ways, of our culturally endorsed alienation from our experiential embodiment. Let us review this history briefly and schematically.

It is well known that many of Freud's contemporaries were quick to dispute the significance of embodied sensuality in the individual's formation. Alfred Adler, who broke with the International Psychoanalytic Association in 1911, was explicit in his distaste for Freud's interest in bodily matters, speculating instead about power motivation (Adler, 1998). Jung's dispute with Freud, which culminated in his 1913 resignation from the International, was more protracted and complex (Freud & Jung, 1994).

There is no doubt that Jung's break with Freud involved, to a significant measure, his difficulties accepting the sexual and sensual nature of unconscious forces (Samuels, Shorter & Plaut, 2003). Yet there are some qualifications that need to be stated here, for Jung's dislike of Freud's ideas on libidinality were quite nuanced, and any assessment of his attitudes toward embodiment is likely to be somewhat mixed.

Jung's seminars on Nietzsche discuss somatic issues favorably and quite extensively (Jarrett, 1997; Jung, 1988). In other works, he explicitly criticized the mind/body dichotomy as "artificial," wrote about the "psychic realm of subtle bodies" as well as the "breath body," and insisted that the symbols of self formation all arise from within our bodily depths. However, it remains true that Jung is less known for these insights than for his important investigations into the transpersonal or archetypal sources of the unconscious (e.g., Jung, 1968a, 1970, 1971, 1981). A general avoidance of the body and its sexual nature can be detected in much of the analytical or archetypal psychology that follows Jung. However, there are exceptions to this, for example in the work of Marion Woodman (Cater, 2005), and a number of Jungian theorists have made some interesting attempts to remedy this avoidance (e.g., Bosnak, 2007; Costello, 2006; Goodchild, 2001; McNeely, 1987; Paris, 2007; Ramos, 2004).

Throughout the early twentieth century, there were repeated attacks on psychoanalysis precisely because of its apparent emphasis on the sexuality of the body. Ironically, this occurred despite the fact that the history of post-Freudian psychoanalytic theorizing may be characterized as a series of conceptual retreats away from somatic psychology (as discussed in Chapter 1). Another aspect of this general retreat was the way in which many of Freud's early followers appeared to maintain an interest in the sexual aspects of the individual's development but still failed to grasp the radical implications of his notion of libidinality. These theorists (including Ernst Jones, Karl Abraham, Anna Freud, and innumerable subsequent contributors) continued to write about the body and its sexual functioning, but increasingly tended to conceptualize it in terms of behavioral activities that might be of great significance in the formation of the individual. As was discussed in Chapter 1, this sort of conceptualization loyally perpetuates a disciplinary emphasis on talking *about* the body, even while comprising a retreat away from Freud's radical discoveries about the formation of the psyche in the energetic experiences of embodiment. The notion of libidinality is preserved in name only, its radical essence lost beneath theories concerning the ego's relationship with various sexual acts. However, such theorizing contributed greatly to the popular vision of psychoanalysis as being "all about sex" and, in a sexually inhibited and prohibitive culture, this vision stirred near hysterical excitement. Here we might consider not only the extent to which the early psychoanalysts were lampooned as "sex crazed," but also the provocative quality of psychoanalysis and its impact on the literati and intelligentsia from the 1920s onwards; examples of which could be found in the work of André Breton, D. H. Lawrence, Henry Miller, and many others.

This two-sided repudiation of Freud's significance as a somatic psychologist is, of course, an oversimplification, and there are some important exceptions to this schematization that need to be mentioned. The theoretical and practical work of three of Freud's contemporaries — Otto Rank, Sándor Ferenczi, and Wilhelm Reich — has had enduring influence on the field of somatic psychology or bodymind therapy, and in particular has contributed to the contemporary development of what I am calling *somatic psychodynamics*.

Before this is discussed, let us additionally mention the work of three other psychoanalysts, who were closely associated with Rank, Ferenczi and Reich: Otto Gross, Georg Groddeck and (somewhat later) Michael Balint.

Gross, who was ostracized for his radicalism and who has virtually been expunged from the annals of psychoanalytic history, was nevertheless a significant influence in understanding the significance of bodily experience (Green, 1999; Gross, 2008). He was a brilliant revolutionary thinker, who championed a countercultural philosophy that questioned Freud's insistence that civilization requires repression, and that advocated the individual's freedom of embodied expression in relation to social norms. He was an ardent anti-authoritarian advocate of repression-free upbringing and of emancipation from patriarchal hierarchies, whose views profoundly impacted Jung and Reich as well as many others. In some respects, Gross' impact in the early years of European psychoanalysis is perhaps comparable to the impact that Herbert Marcuse would have on social thought in the latter part of the twentieth century.

Groddeck, who is credited by Freud (1923a) with naming the unconscious as *Es* or "It" (later "Id"), was an important pioneer in psychosomatic thinking (Groddeck, 1961, 1988; Rudnytsky, 2003). Indeed, he has been called the "father" of psychosomatic medicine. He treated chronically ill patients, often introducing them to naturopathic treatments alongside their psychoanalytic therapy. It has been humorously said that, when patients would come to Groddeck for psychoanalysis, he would offer them massage, and when they would come to him for massage, he would offer them psychoanalysis! This perhaps illustrates Groddeck's radicalism and his place alongside Rank, Ferenczi, and Reich in understanding the bodymind context of healing practices.

Finally, although coming later in psychoanalytic history, the work of Michael Balint (1987, 1995), who was a student of Ferenczi's, is important here because it was Balint who remained within the International, yet managed to develop some of the pioneering ideas on the significance of much of what had been developed by Rank, Ferenczi, Reich, Gross and Groddeck. Let us look more closely at the work of the first three of these contributors.

Rank was a brilliant thinker with wide-ranging interests in art, myth, religion, and philosophy, as well as psychotherapy. He worked closely with Freud from 1905 and was one of his earliest, most prolific, and favored collaborators. However, his 1924 publication, *The Trauma of Birth*, which is perhaps his work of most enduring influence, precipitated a rift with Freud that eventually led to his resignation from Freud's inner circle (Rank, 1994). Rank argued that the psychological constitution of the human condition was significantly influenced by events prior to the Oedipal complex; it was implied that the impact of such events, occurring prior to the development of language, would entail their bodily inscription. Freud, however, opposed any appearance of theoretical notions that might diminish his theoretical emphasis on the Oedipal complex. The latter is a universal conflict involving mental representations (thoughts and feelings about the subject's self in relation to two other persons, typically one male and one female, and the subject's interpretation of what might occur between them). In 1925, Rank was the first to refer to "pre-Oedipal" conflicts in a public lecture, and subsequently the rift deepened irrevocably (Rank, 1996).

The importance of Oedipal complexities in the constitution of every human being is perhaps undeniable (Barratt, 2009b). These universal conflicts involve the comparatively sophisticated processing of ambivalent feelings. In the traditional nuclear family, these usually involve the child's representations of maternal and paternal figures, as well as the perceived or fantasized relationship between them. Aspects of the individual's Oedipal complex may often be inscribed in the body, but typically the ambivalent and conflicting emotions that are involved in this complex are potentially available for linguistic formulation even as they impact bodily structures and functions. By contrast, the individual's primary passage through pre-Oedipal conflicts occurs prior to the acquisition of language. Birth trauma, for example, which refers to the impact of the passage down the birth canal, can only be impressed upon the individual in a way that is holistic or somatic. It is very unlikely that the event is available for an intellectualized cognition that is dissociated from bodily experience! In general, the emotions involved in pre-Oedipal conflicts are usually less nuanced than those of the Oedipal complex. Thus they are invariably impressed upon, and expressed by, the child in bodily modalities. The pre-Oedipal conflicts of the pre-linguistic years may subsequently be crudely translated into language, in a process akin to Jung's notion of retrospective fantasy, but they remain less accessible to memory that is dissociated from bodily experience. This then is central to Rank's importance in the unfolding of somatic psychology.

However, Rank's influence on the emergence of bodymind therapy goes further. Rank pioneered an experiential, here-and-now approach to therapy, which is congruent with his interest in pre-Oedipal aspects of psychological suffering. After all, it may be possible to address Oedipal issues with clinical methods that remain "in the head" (although the practical success of this is likely to be limited). But it is, in principle, impossible to address pre-Oedipal anguish without a therapeutic procedure that involves listening to the voicing of embodied experience. In this respect, Rank had a considerable influence on leading figures in the emergence of bodymind therapy such as Paul Goodman, who co-founded gestalt therapy with Fritz Perls as well as others, and Stanislav Grof, whose innovative work on primal integration and holotropic breathwork was mentioned in Chapter 5. Rank also influenced other major figures in the development of psychological theories that were not particularly interested in the development of bodymind therapies (e.g., Becker, 1998; Fox, 2002; Rogers, 1989), as well as a generation of writers and artists such as Anaïs Nin and Henry Miller.

At the beginning of their psychoanalytic careers in the early years of the twentieth century, Ferenczi and Rank were collaborators (Ferenczi & Rank, 1956). Although Ferenczi's influence on the later emergence of somatic psychology is less direct and less well known than Rank's, the nature of his dissent from the orthodoxy of the major post-1914 schools of psychoanalysis is important. Ferenczi championed more active methods of intervention that included physical contact with his patients as well as an emphasis on the here-and-now of their embodied experience (Aron & Harris, 1993; Ferenczi, 2008). He also advised therapists to attend to their own inner experiences (sometimes pejoratively called "countertransference"), and not just to the logic of their observations and inferences about their patient's functioning. He thus foreshadowed later psychoanalytic work that would stress the centrality of interaction, intimacy, and mutuality in the therapeutic setting (e.g., Ehrenberg, 1992). Ferenczi contributed further to the development of somatic psychology with his remarkable theory of human genitality

(Ferenczi, 1989). His 1938 book, *Thalassa*, was titled in honor of the primordial sea goddess of Greek mythology, who was supposedly the mother of Aphrodite, who figures in the tale of Eros and Psyche, and who was both Adonis' lover and his surrogate mother. It is a visionary expansion of Freud's theories of libidinality as a speculative account of the most basic human urges and the foundation of our experience in the natural resources of our sexual embodiment.

Reich's legacy, as the "father" of somatic psychology, has already been discussed in Chapter 5. Here we need to note briefly how his energetic approach to the human condition evolved over his lifetime, and we need to focus on the questions raised for somatic psychodynamics by this evolution. Reich began his career emphasizing and developing a strictly Freudian view of libidinality. He was vehemently opposed to the forces of sexual repression and oppression, both intrapsychic and sociocultural, and he wrote extensively about the harm done by the blocking of libidinal energy. His theory of the character armoring, by which the patient's potential for free-flowing energy, exemplified by orgasmic processes, is blocked, and thus the patient's fear of this energy is expressed, is as valuable for the clinician today as it was when *Character Analysis* was finally published in 1933 (Reich, 1980a). Often this theory is depicted as an effort to take psychoanalysis from the treatment of neuroses into the treatment of character problems. In some respects this is correct. But Reich's work on character armoring entailed the need for clinical methods that go beyond "just talking" — if indeed "just talking" implies the telling and retelling of the patient's stories, rather than a renewed practice of listening to the voicing of our embodied experience. In this respect, it was already a deviation from what had, by the 1930s, become psychoanalytic orthodoxy. Reich began to modify the procedures of psychoanalysis, working intentionally with the patient's breathing, touching the patient in order to facilitate this work, and having patient's adjust their posture or move around the consulting room so as to identify their character armoring. This approach to the dissolution of character armoring was notably confrontational, and Reich's methods of intervention quite direct.

As Reich's work progressed, he elaborated the notion of libidinal energy into his distinctive theory of *orgone energy*. This elaboration is also characterized by some commentators as a break with psychoanalysis, and we will shortly evaluate this opinion. The shift to orgone theory is signaled in Reich's 1927 book, *The Function of the Orgasm* (Reich, 1986). Whereas libidinality had been mostly understood as being contained within the confines of the body (although exchanged and perhaps enhanced by intimate activities such as intercourse), orgone was understood to pervade the universe, and thus its mobilization might be beneficial for activities as seemingly diverse as cancer treatment and inducing rainfall. As is well known, Reich started to develop techniques for accumulating orgone energy, such as a box for patients with various ailments to sit in so that they might be healed. By 1934, Reich began a series of experiments with "bions" — elementary vesicles held to contain the lifeforce called orgone. His 1938 book, *The Bion Experiments on the Origin of Life*, described these experiments (Reich, 1979). By 1940 Reich was engaged in an unsuccessful debate with Albert Einstein over the validity of his results. As was discussed in Chapter 5, Reich's claims concerning the curative properties of orgone energy were widely contested and resulted in his imprisonment.

Reich's revision of libidinality into the theory of orgone is presented as a necessary recognition that this energy, if it exists at all (and there are many who doubt its existence), is not confined to the human body. The kinetic flow of libidinality or orgone is both material and immaterial, neither meaningless nor a meaning that can be represented, and is thus deeply mysterious or even incomprehensible to the tradition of Western rationality. As such, it pervades the universe just as *prāṇā*, , or Spirit are known to do. According to Feuerstein (1998a, p. 466), this was also called *mana* by the Polynesians, *orenda* by the native Americans, and was know in ancient Germanic cultures as *od*. The lifeforce is to be found in the world around the human body and not just contained within it. The correspondence between Reich's ideas and the insights of the Dharmic tradition (Vedic, Buddhist, Taoist, and so forth) is striking; yet there seems to be little evidence that Reich was aware of this correspondence or had any significant familiarity with writings from that tradition (cf., Sharaf, 1994).

It seems clear that Reich's refashioning of libidinal theory also heralded a shift in which he became a little less interested in individual treatment modalities and a little more interested in experimental and speculative study of the origins of life itself. Indeed, some might argue that he moved away from psychology as such; for example, in the preface to the third edition of *Character Analysis*, he wrote that "in orgone therapy, we proceed *bio-energetically* and no longer psychologically" (1949, p. ix). However, the critical question is not so much whether Reich became disinterested in psychology, but whether he moved away from a psychodynamic approach to one that is linear or univocal, and thus ultimately limited the value of his perspectives for the emergence of somatic psychology and bodymind therapy.

Let us consider here what *psychodynamics* might mean. There seem to be three characteristics of a psychodynamic approach to the study of the human condition:

- This approach is concerned with *meaningfulness*, including the inner meanings that events hold for the individual, and including the meaning of matters that appear to have been rendered meaningless.
- This approach recognizes that meanings are in a perpetual condition of *movement* and *contradictoriness* or conflict; and that any apparent resolution to such conflict is, at most, temporarily static or reified.
- This approach acknowledges that the meaning of things is always multiple, *interdependent* and *nonlinear* or, under a different terminology, *polysemous* (implying that multiple and even contradictory meanings can be conveyed by a single semiotic event); thus, nothing can be properly understood if reduced to a linear conflict between two forces.

In this regard, it can be seen that Freud's discovery of the repressed unconscious was, from the very start, psychodynamic, because it was not merely the discovery of some region of ideas and feelings outside of consciousness. Rather, it was the discovery that consciousness perpetually reveals and conceals meaningfulness that is otherwise than that which it takes itself to mean (Barratt, 1993). Likewise it can be seen that the metatheoretical platform of behaviorist psychology, along with some of the major schools of psychoanalysis since Freud (as was discussed in Chapter 1), eliminate the possibility of psychodynamic insight into the functioning of the human condition.

Psychodynamic insight suggests that any theory positing a linear conflict between two forces is a simplification that commits an ideological distortion. For example, a theory that depicts the individual as a unified being in relation to a univocal exteriority or world, even if this relationship is seen as bidirectional (in which the world influences the formation of the individual and, to some degree, the individual influences the construction of his or her world) commits itself to the ideological erasure of the psychodynamic reality. By contrast, a psychodynamic appreciation involves complexity. It is an appreciation of reality in terms of its interdependent nonlinear dynamics (Cowan, Pines & Meltzer, 1999; Kauffman, 1996, 2002, 2008; Morin, 2008).

In this context, it can be argued that there is a serious inherent limitation or flaw in any theoretical position that merely points to the lifeforce and the forces that obstruct the freedom of its flow, or that tends to depict the individual's interior vitality heroically struggling against the exterior forces of repression and oppression. I have referred to this error as an ideological distortion, because it presents matters in such a manner as to appear resolvable without a foundational reworking of the forces that originally established the conflict. In this sense, an oversimplification is committed that constricts the possibility of genuinely emancipative momentum. In short, such theories lose their potential for psychodynamic sophistication, and it often seems that Reich's theoretical work is dangerously close to such simplification. Reich frequently drifts away from psychodynamics toward reductive theoretical position, *not* because his understanding of energy was flawed, but rather because he only had a Newtonian language in which to describe it. Reich remains centrally significant in the emergence of somatic psychology and bodymind therapy. Yet as we progress through the twenty-first century, there are many aspects of his work and of the work of those in his lineage that need to be surpassed.

8

Philosophical and Cultural Studies

The impetus for the contemporary emergence of somatic psychology and bodymind therapy does not come only from within the field of psychology. Rather, there are several entwined developments in twentieth century philosophy and cultural studies that have contributed strongly to this momentum and that contextualize its advance. One critical scheme by which to consider these developments is as follows. Every endeavor of human inquiry has a *subject matter*, a *method* for studying that subject matter, and an *ethical-political context* of forces that create an interest in the subject matter and method to be pursued. The notion of "interests" that animate any pursuit of knowledge has been advanced by Jürgen Habermas (1972), and constitutes this ideological dimension of any investigation (cf., Teo, 2005). The utility of this scheme is that it enables us to examine how subject matters and methods are selected and defined, and how ideological forces impact these processes of selection and definition. In relation to twentieth century philosophy and cultural studies, this chapter will briefly review how a hundred years of deliberation has set the stage for a (re)turn to the experience of embodiment as the essential starting-point and the necessary center of any scientific study of the human psyche.

Scholarly philosophy, at the beginning of the twentieth century, was at several crossroads (as was suggested in Chapter 2). Much of the field, at least in Europe, was still reeling from the grandly synthetic, yet somewhat abstract, achievements of Hegel's 1805 *Phenomenology* (Hegel, 1977). Søren Kierkegaard, who stands at the head of the existentialist tradition, suggested that Hegel had omitted the concrete experience of individual existence. Karl Marx argued that Hegel's synthesis was an abstraction that obscured the significance of the forces of material production in the development of human consciousness and societal formations. Friedrich Nietzsche opposed Hegelian philosophy on grounds that might be called ethical and aesthetic, and his own work included some dazzling discussions of the earth and of human embodiment. None of these thinkers were overly focused on epistemological questions — the formal exploration of how we know things — which had been the focus of pre-Hegelian philosophers such as Descartes and Kant. In the early years of the twentieth century, much of the field of philosophy was still under the sway of the Cartesian-Kantian lineage, and it is to this lineage that Husserlian phenomenology is a response.

Edmund Husserl aimed to deepen and extend Cartesian reflections on the subject of experience and understanding, and thus to resolve what he saw as the crisis of natural-scientific objectivism (as well as the naïve objectivism associated with it). He described the crisis as deriving from the scientific ideals of the modern era (which involve the Galilean mathematization of nature and the generalization of Euclidean geometry into a formal and universally applicable mathematics). Although modern science, while operating under these ideals, produces impressive technological successes, Husserl argued that it eclipses the concrete experience of the subject's constitution and, because the subjective foundation of the natural attitude (the subject-object split assumed by modern science) was ignored or naïvely assumed, modern science is in danger of losing the meaningfulness of the objectivist world it studies (Husserl, 1970, 1974). Any discipline that assumes the natural-scientific attitude can wind up losing the meaningfulness of the world for humanity, promoting a sort of pseudo-rationality that can say little or nothing about the creative subject.

Like Freud, Husserl was a student of Brentano (1995a, 1995b, 2001), and a contemporary of Bergson (2007a, 2007b, 2008). Following Brentano, Husserl took intentionality to be the primary characteristic of consciousness — mental events are directed toward something as a sort of striving, and are in this sense inherently purposive. Thus, he understood that mental events are presented and have meaning existentially before they are judged predicatively (in an established subject-object framework). To side-step the natural-scientific attitude, and thus to overcome the crisis of objectivistic epistemology, Husserl developed phenomenology as a transcendental and reflective method that inspects or intuits the essential structures of a "purified consciousness" that is freed from the suppositions of the natural attitude (Husserl, 1960, 1969). Husserl aimed to establish an "egology" without ontology — a universal science of ultimate foundations, or original meaning structures — that would investigate how the world is known, without making assumptions about the existence or the nature of the world that is to be known.

Here we do not need to detail the methodical practices of Husserl's phenomenology — his procedures of bracketing, "reduction" or *epoché*. Rather, we must both highlight the importance of the endeavor as a rigorous return to the structures of consciousness or experience, and at the same time point out how quickly Husserl's reflective investigations became abstractly transcendental and thus distanced from the embodied nature of experience. Even Husserl's earliest efforts at the descriptive phenomenology of empirical consciousness illustrate this limitation, and the later investigations of transcendental phenomenology address only an "anonymously functioning subjectivity." For example, Husserl's studies between 1904 and 1910, which were gathered together as *The Phenomenology of Internal Time-Consciousness* (Husserl 1964), and which are arguably the best exemplar of Husserlian procedures, are both highly illuminating and notably disembodied.

While it was left to Paul Ricoeur (1967, 1974) and others to criticize the Husserlian program for the ahistorical character of its investigations (and also for ignoring the significance of the cultural context of thinking), it is principally Maurice Merleau-Ponty who, as a student of Husserl's, takes up the crucial issue of the phenomenology of embodiment and, in so doing, provides a critique of the disembodied character of the Husserlian egology and of the entire Cartesian-Kantian tradition. In many respects, Merleau-Ponty's philosophy attempts to use the phenomenological method to answer ontological questions (questions about the nature and relations of being), and in this his endeavor parallels those of Heidegger (1962, 1972) and Jean-Paul Sartre (1956, 1965).

From his earliest work, Merleau-Ponty (1962, 1963) was troubled by the abstract emptiness of the Cartesian-Kantian subject or ego — the subjective side of the natural attitude that sustains the thinking and speaking of the modern episteme — an emptiness that a later commentator would characterize as a "black hole" (Kolakowski, 1988). He argues for the primacy of perception in how we experience and engage in the world, and thus for the incarnate nature of subjectivity (Merleau-Ponty, 1964, 1968, 1973b). Going beyond the Husserlian distinction between the acts of thinking and the intentional objects of thought, Merleau-Ponty shows how this duality misses the foundational properties of embodiment as providing a "preconscious" and pre-predicative "prehension" or "grasp" (in French, *prise*) toward itself and its world. It also misses crucial aspects of our experience of temporality and the "other" (as developed so well in the work of Emmanuel Levinas). Taking the investigation of perception as his starting-point, Merleau-Ponty argues that although one's own body (*le corps propre*) can be treated as a thing — conceptually represented and treated as the instrumental object of scientific or quasi-scientific scrutiny. It is also the perennial condition of all experience, the perpetual condition of our openness to our world and to our belongingness within it. Our corporeality is intrinsically intentional and expressive, and is thus the foundation for a subject or ego that can perceive and cognize itself and its world (cf., Kleinberg-Levin, 1985, 2009).

In this way, Merleau-Ponty's phenomenology discovers that our carnality (or "chiasmic flesh" as he sometimes calls it) expresses a special mode of intentionality, an essential unity of flesh, that is anterior to and founding of the possibility of subjects and objects, the acquisition of language, and the formation of concepts about the body, the self, and the world around us (Merleau-Ponty, 1968, 1973a, 1973b; Toadvine, 2009). In this respect, Merleau-Ponty elaborates the insights of Brentano and Freud suggesting that our experience of presence is first and foremost a bodily experience (Brentano, 1995a, 1995b; Freud, 1925). Our corporeality is an inherent consciousness that grounds the possibility of all other — egological — formations of consciousness, and is thus the foundational nature of *psyche*.

Although it has been argued elsewhere that Merleau-Ponty's appreciation of the notion of libidinality is limited (Barratt, 1993, pp. 144–149), the importance of Merleau-Ponty's redirection and development of phenomenology for the emergence of somatic psychology and bodymind therapy is evident (cf., Spiegelberg, 1972; Zaner, 1964). His philosophy has also had considerable influence on the recent advocacy of what has been called "anti-cognitivist cognitive science" and its opposition to intellectualist psychology. For example, Merleau-Ponty's influence is seen in Dreyfus' seminal critique of the computational account of mental functioning or "cognitivism." Dreyfus offers a connectionist argument that our corporeal "know-how" precedes or grounds, and is not reducible to, the operation of discrete or independent representational procedures in our perceptual and conceptual consciousness (Dreyfus, 1972, 1992; Dreyfus & Dreyfus, 1992; Todes, 2001). This critique has supported the further development of existential phenomenology and the recent germination of "neurophenomenology" (cf., Clark, 1998, 2008; Gallagher, 2005; Noë, 2006, 2009; Petitot, Varela, Pachoud & Roy, 2000; Varela, Thompson & Rosch, 1992). We will revisit some of these issues in Chapter 12.

Subsequent to the work of European philosophers such as Husserl and Merleau-Ponty, the methods of "phenomenology" have blossomed in North American psychology, particularly with the work of Giorgi (1970, 1985) and others. In the past few decades, qualitative methods in psychology — heirs to Brentano's manifesto for a descriptive psychology — have gained respectability and popularity (e.g., Camic, Rhodes & Yardley, 2003). However, many of these methods owe more to the nineteenth century philosophical traditions of Friedrich Schleiermacher's and Wilhelm Dilthey's descriptive hermeneutics than they do to the rigors of twentieth century phenomenology and hermeneutic ontology (cf., Barratt, 1984). Nevertheless, the momentum of these shifts in methodology is clearly hospitable to the further emergence of somatic psychology and bodymind therapy.

Alongside these developments in philosophy — and in the methodologies of human inquiry — there has been, in the past several decades, a powerful upsurge of interest in the topics of embodiment and the disposition of bodies within the field of cultural studies and in the social sciences in general (e.g., Foucault, 1988–1990; Weiss & Haber, 1999). For example, poststructuralist theorists, such as Foucault, Jacques Derrida, Georges Bataille, Julia Kristeva, and Roland Barthes, attempted to focus on the body, in an effort to counteract the somewhat abstract or disembodied structuralist thinking that so greatly influenced the social sciences following the 1916 publication of Ferdinand de Saussure's lectures in linguistics. More recently, the sociology of the body has become a popular branch of the discipline, although much of the better literature in this field has been done by French scholars and remains unavailable in translation. Examples of this are the voluminous work of Gleyse (1997) on the history of the body's utilization, and of scholars such as Vigarello (2005) on other aspects of the history of the body. Some of the interest in this field is in the sociology of health and illness (e.g., Conrad, 2008; Nettleton, 2006), in medical sociology (Cockerham, 2006), and in the cross-cultural findings of medical anthropology (e.g., Scheper-Hughes & Wacquant, 2003). Although the disposition of bodies has been a central topic in cultural anthropology since the nineteenth century (e.g., Blacking, 1977), it is only toward the end of the twentieth century that it came to receive more attention as central to any social theory (e.g., Berman, 1989; Featherstone, Hepworth & Turner, 1991; Shilling, 2003; Turner, 2000, 2008) and of profound importance to the entire field of sociology (e.g., Feher, Nadaff & Tazi, 1989; Fraser, 2005; Polhemus, 1978; Scott, 1993). Some of the best literature in this area has been feminist in its orientation (e.g., Price & Shildrick, 1999), and of particular importance are the psychoanalytic writings of Julia Kristeva, Luce Irigaray, and others (e.g., Cixous & Clément, 1986; Gallop, 1988; Irigaray, 1985a, 1985b; Kristeva, 1980, 1987). The issue that must be taken up here is as follows. The literature on the anthropology and sociology of the body is focused on studying our representations of the body as well as the social uses of the body, and this is crucially different from a study of our experience of embodiment (e.g., Berman, 1989; Lock & Farquhar, 2007). It was suggested in Chapters 1 and 5 that somatic psychology might be a psychology the body, rather than one that is *about* the body or that directs its activities *at* the body. So while there is certainly interest in the researches of disciplines such as the anthropology or sociology of the body, psychosomatic medicine (in the frame of modern allopathic science), rehabilitation medicine, sports psychology, human factors engineering, and so forth, there is also a sense in which somatic psychology and bodymind therapy might be radically different from these enterprises.

As stated earlier, the prepositional distinction between a psychology the body, and a discipline that is *about* the body or a procedure that is directed *at* the body intimates a profound difference in the spatio-temporal or ontological relations that are assumed or engaged in the course of the inquiry. Even our very brief mention of some of the ideas of Brentano, Freud, Husserl and Merleau-Ponty helps us appreciate this difference. For there is a critical point here that has at least three aspects:

- Embodied experience is an experience of *presence* (or more accurately of absenting and presencing) that is anterior to, and foundational of, the subject-object dualism, and our ability to represent things as self and other. In Merleau-Ponty's terms, our fleshly incarnate subjectivity comprises a pre-predicative "prehension" or "grasping" toward itself and its world that is prior to the constitution of self and world (that is, prior to the representational formation of subject and object). This is parallel to what Freud was struggling to articulate when he argued that we experience the qualities of things — libidinally — before we judge whether they exist or not. It is also parallel to Brentano's argument that judgment may be pre-predicative and that there is, so to speak, an experience of presentation (or absencing and presencing) on which predicative judgment (the judgment made by a subject about an object) depends, but which is prior to such predication (and the subject-object dualism). Recall here the argument that mental events are always intentional in that they are directed towards, or strive towards, something even before the subject-object framework is established.

- If consciousness is indeed intentional, then it must be added that the entire body is a mode of consciousness (even while being a consciousness that is often repressed or "preconscious" with respect to the consciousness of our mental representations). This is the consciousness of libidinality, or fleshly energy that is "in" but not "of" our mental constructs. However, the reflective or secondary consciousness of our representations is, according to psychoanalytic discoveries, alienated from the consciousness of our corporeality.

- We can treat the body as a thing, a machine, and as something "other" that just happens to be in our possession. For instance, we can mentally represent our hands, feet, intestines and genitals, for we have concepts of them and, at least to some degree, these concepts can be used to govern the objects they represent (e.g., I think of moving my hand, and then my hand moves). This is an *instrumental* treatment of our embodiment, and it perpetuates the *conceptual-objectivistic alienation* of our body as other. However, we can also listen to the presentations or presencing of our carnal subjectivity. This engages what might be called an *evocative-integrative treatment of our embodied being-becoming* that is profoundly different from, and usually incompatible with, an instrumental treatment of the body. Here our alienation from our embodied experience is overcome, although a sort of dialogic estrangement inevitably continues, as the body is mobilized as an interlocutor to whom we belong. This principle is crucial to the viability of bodymind therapy.

If these three arguments express the truthfulness of the human condition, then it can be seen that disciplines that address the body in an objectivistic manner, that attempt to represent an improved understanding of how we represent the body or how we use it, and procedures that are designed to improve the performance of the body, are all significantly divergent from what might be the epistemology and ontology of somatic psychology and bodymind therapy. Nevertheless, the significance of at least two developments in the human sciences must be noted. Although it may be objectivistic, the rise of qualitative methods in psychology suggests a return to the notion that the study of the psyche should be a study of experience. The rise of interest in the body on the part of such diverse disciplines as anthropology, sociology and cultural studies, also indicates an important shift. Both developments entail a scholarly momentum that is conducive to, and influenced by, the emergence of somatic psychology and bodymind therapy (e.g., Csordas, 1994).

As indicated previously, the twentieth century's story of shifts in philosophy and cultural studies illustrates the extent to which the discipline of psychology has frequently established peculiar and unwarranted priorities between its subject matter, its methods, and its ethical-political impetus. In short, the method of investigation has often determined the subject matter to be investigated, and political forces have often determined both method and subject matter even while these lines of influence are disguised under the rationale of commitment to the ideals of a natural-scientific approach. Brief examples from the academic and the applied realms will suffice to illustrate this ubiquitous tendency.

In academic psychology, the notion of intelligence has played a central role in the accomplishments of psychological science through the twentieth century. While it is obvious that different people have different abilities, the veracity of the construct of *intelligence* is far from obvious. People do not experience themselves as intelligent. Rather, they may experience themselves as able to do this or that. Thus, the idea that there is such a thing as intelligence (even if one breaks it down into verbal intelligence, mathematical intelligence, emotional intelligence, and so forth) is far from inevitable on the basis of experience. Moreover, it would be hard to sustain the argument that the construct of intelligence serves the individual's need for healing or that it directly facilitates emancipative activities. The construct is generated politically and the history of the notion amply exhibits its political agenda. The construct serves a societal ambition to discriminate individuals and groups of individuals so that they can be placed more effectively in the workplace and the social order. While such discrimination may sometimes have benign consequences, in general, the ethicality of the procedure is questionable. When one speaks of societal ambition in this manner, what is usually implied is the ambition of the dominant social groups; and when one speaks of the effectiveness of a procedure by which people are socially positioned, the criteria usually entail the economics of the profit motive. The politically driven construct then takes off not because of its subject matter, but because it can be measured and the data utilized for political and socioeconomic ends. In this way, the method drives the subject matter, and the investments made by political forces determine the utility of the method. The entire program is rationalized by its natural-scientific cloaking.

A further illustration of this would be the recent idealization of any method for changing individual behavior that is *evidence-based*. For a clinical treatment to be evidence-based implies that it has met objective criteria of effectiveness, under reasonably rigorous conditions of experimental testing. Such criteria have to be behavioral, since that is what can be measured — after all, evidence comes from the Latin *evidere*, to make visible, and it is assumed that what can be made visible must be quantitatively assessable. The effectiveness of a clinical treatment usually refers to the individual's behavioral adaptation or adjustment to the circumstances of his or her social and cultural position. Again, much of this procedure may have benign consequences, but the assumptions on which it is based are designed to support the economic and political interests of the dominant social order. Evidence-based assessments miss much of the human experience, and much of the human experience is ideologically distorted by their operation. It would be hard to sustain the argument that evidence-based procedures are responsive to the experience of the individual — although presumptive claims about the implications that behavior must have for inner experience are often made. Evidence-based procedures cater to the agenda of adjusting individual behavior to fit the existing social order. It can hardly be claimed that they place a premium on experience, or that their operation is a healing of the psyche, let alone directed at the liberation of the individual's potential.

The significance of the emergence of somatic psychology is that it establishes, or re-establishes, human experience as the primary subject matter of any inquiry into the psyche and that it acknowledges the primacy of embodied experience. Unlike much twentieth century psychology, the subject matter determines the method of inquiry, and not vice versa. And unlike most of the proceedings of this discipline through the twentieth century, somatic psychology follows ethical and political principles that might be called emancipative (as will be further discussed in Chapter 15). The increased attention that is being paid to the body in the objectivistic researches of anthropology, sociology, medicine and other disciplines, is not congruent with the characteristics of somatic psychology, but has certainly provided a scholarly context within which somatic psychology has begun to make its mark.

Western Traditions of Bodywork

A considerable history of bodywork methods that have steadily gained popularity in North America and Europe also provides a context for the emergence of somatic psychology and especially for the escalating interest in bodymind therapeutic modalities. Here the term "bodywork" refers to any physical manipulation intended to facilitate healing. In this chapter, we will briefly review some of the salient material on the history of such therapies, including both massage and movement methodologies, and then proceed to a short discussion of the role of bodily awareness and appreciation, which defines the sort of bodywork that is now often called *somatics*.

Although massage and other physically manipulative treatments have been known for thousands of years, their evolution in the twentieth century is concurrent with the history of osteopathic and chiropractic healthcare. Osteopathy was founded at the end of the nineteenth century by Andrew Taylor Still, largely as a protest against the non-holistic and drug-dependent practices of allopathic medicine (Still, 1992, 2009). Still was also critical of the allopathic tendency to treat symptoms rather than to address the causes of disease. Osteopathy started in the United States and was established in Europe in the early decades of the twentieth century. Its practice has grown quite steadily.

Among the distinctive features of osteopathic healthcare is the practice of osteopathic manipulative medicine, which is a manual therapy that addresses the muscular-skeletal system to resolve what are called "somatic dysfunctions" by facilitating the body's own recuperative faculties (Ward, Hruby, Jerome, Jones & Kappler, 2002). Osteopathy operates on eight principles or empirical laws, which are widely applicable to many modalities of bodywork: (1) the body is a unit; (2) structure and function are reciprocally interrelated; (3) the body possesses self-regulatory mechanisms; (4) the body has the inherent capacity to defend and repair itself; (5) when the normal adaptability is disrupted, or when environmental changes overcome the body's capacity for self-maintenance, disease may ensue; (6) the movement of body fluids is essential to the maintenance of health; (7) the nerves play a crucial part in controlling the fluids of the body; (8) there are somatic components to disease that are not only manifestations of disease, but also are factors that contribute to maintenance of the disease state. In these principles, an emphasis on what might be called the *wisdom of the body* is evident. This perhaps contrasts with healthcare practices that tend to frame the body as an antagonistic liability, which is concordant with the treatment of the body as a machine-like object to be investigated and used instrumentally. This is perhaps why — at least in Europe — the history of osteopathy has often been associated with the increased interest in complementary and alternative medical practices such as homeopathy and naturopathy.

Chiropractic manipulative treatments were also initiated in the United States at the end of the nineteenth century by Daniel David Palmer (Haldeman, 2004; Palmer, 2006). These treatments are usually focused on spinal manipulation, and emphasize the effects of this procedure on the nervous system. Chiropractic healthcare follows at least four principles: (1) holism; (2) the avoidance, whenever possible, of invasive procedures or use of medication; (3) belief in the body's innate intelligence or self-healing capacity (body wisdom), which is its ability to regulate itself by homeostasis if misalignments are adjusted; (4) a patient-centered approach in which patients are ultimately considered responsible for their health. Despite legal attacks by allopathic organizations, the chiropractic profession has generally flourished through the twentieth century and is now the most widespread form of complementary or alternative healthcare, other than the general practice of massage.

Whereas osteopathic and chiropractic interventions address the hard tissues of the body (bones, joints, spinal alignment, and so forth), massage practices generally focus on the soft tissues (muscles, tendons, ligaments, skin, tissue around the joints, and other connective tissues). Massage has probably always been used informally for the relief of tension and for pleasure (Calvert, 2002). It has a longstanding history of practice in Europe and elsewhere, but was formally introduced into the United States by physicians in the mid-1800s. These professionals were impressed by the methods propagated by the Swedish medical-gymnastic practitioner, Pehr Henrik Ling, and saw his methods as an important adjunct to allopathic practice. By the twentieth century, massage was regularly prescribed by physicians for a range of conditions. This usage declined somewhat in the 1930s and 1940s — decades in which there were numerous pharmacological innovations available to allopathic practitioners. The popularity of massage then boomed after the 1960s not only because of the expansion of sports medicine, but also because of the increasing public interest in complementary and alternative healthcare methods. Today massage is the most utilized of all these methods (Calvert, 2002; Graham, 2008).

Over eighty different modalities of massage have been enumerated. The best known of these is Swedish massage, which has about a half-dozen techniques of touching the soft tissues, and is famous for its style of long flowing strokes. There are, however, many other modalities, including some which have been developed as a result of the influx of Asian methods of healing (see Chapter 10). The definitional commonality of all these modalities is the manipulation of the malleable tissues, often with the goal of relieving muscular tension and increasing the circulation of the blood. Massage may also be directed toward the lymphatic or gastrointestinal systems. It is not only known to relieve muscular pain, but also to be comforting, and to induce sleep. It reduces anxiety, lowers blood pressure and heart rate, and alleviates subclinical depression. It reduces the level of stress hormones and is believed to support the immune system by increasing peripheral blood lymphocytes. It is very likely that there are many other beneficial effects that are less well documented (Fritz & Grosenbach, 2008).

The passivity of the recipient in massage therapy contrasts this method with other Western traditions of bodywork, notably those involving movement (cf., Johnson, 2001). There is a strong western tradition of movement arts and healing practices, including dance practices which extend well beyond the scope of professional choreography (Knaster, 1996). Bodily movement has long been known in the western world as enhancing physical, and hence mental, health, particularly when combined with attention to the breath — improving cardiovascular functioning, muscular-skeletal flexibility and strength (Calais-Germain, 2006, 2007). In the performing arts, however, "modern dance" started in the late nineteenth and early twentieth centuries, both in Europe and in the United States. One aspect of its initial impetus was as a dissent from — and even a protest against — the restrictive forms of traditional dance choreography and ballet (Anderson, 1993; Au, 2002; Reynolds & McCormick, 2003). Somewhat influenced by Konstantin Stanislavski's approach to acting, free-form styles of dance blossomed through the twentieth century. In an important sense, they became politicized as an art of protest against the constriction of the human spirit. For example, free dance and other forms, such as swing dance, were engaged to express resistance to Nazi authoritarianism.

Although the pioneers in this field are not all directly connected with the rise of somatic psychology and bodymind therapy, all of them have had an influence on the cultural context that nurtured this emergence. For example, in Europe, by the mid-twentieth century, the works of Rudolf Laban, Emile Jacques-Dalcroze (who developed eurhythmics), and François Delsarte were influential in the development of this sort of expressionism in dance. Later influences included the innovations of Kurt Jooss, Mary Wigman (who were both Laban's students), and Harald Kreutzberg. Figures such as Loie Fuller, Isadora Duncan and Ruth St. Denis were, in various ways, somewhat influential in spurring similar developments in the United States. St. Denis, in particular, worked with students such as Ted Shawn, Doris Humphrey, Charles Wiedman and Martha Graham, who in turn trained an entire generation of leaders in the field of *avant garde* dance. A slight anomaly in this history is the work of Lester Horton, who refused to move from the west coast to New York (which was the epicenter of dance innovation in North America). Horton incorporated much of the strong and distinguished Native American dance tradition into his work; he was the teacher of several significant innovators in dance such as Alvin Ailey (Perce, Forsythe & Ball, 1992). By the latter half of the twentieth century, the contemporary dance movement in the United States had blossomed spectacularly and continues to this day.

The importance of this history is that the development of free-form, expressionist dance constitutes one aspect of the cultural context in which an interest in the body's inherent capacity for self-expression might flourish. Importantly, this development also leads directly to an appreciation of the potential that awareness of movement has for bodymind healing. Methods that facilitate kinesthetic awareness have become central to the rise of bodymind therapies. As early as 1928, Laban starting developing a system for the study of movements, which later became known as "Labanalysis" or Laban Movement Analysis (Davies, 2006). His student, Irmgard Bartenieff, who worked extensively with polio patients, expanded this method in a system called "Bartenieff Fundamentals," which became influential in North American therapeutics (Bartenieff, 1980). Marion Chace, who studied dance with St. Denis and Shawn, is generally considered one of the main founders of dance therapy in North America, started working in this modality in the 1940s. Chace spent much of her career practicing professionally at

Chestnut Lodge, which was a cutting-edge psychiatric institution, as well as other hospitals. In the 1960s, she started training dance therapists and by 1966 had formed the *American Dance Therapy Association*. However, it was not until the 1970s and 1980s that this method was recognized as a form of psychotherapy, and it has grown steadily in its influence since then (Lewis, 1994; Meekums, 2002; Payne, 2006; Stanton-Jones, 1992).

Dance movement therapy aims to restore to individuals a holistic sense of themselves, and is organized around six principles: (1) body and mind interact, so that a change in movement will affect total functioning; (2) movement reflects personality; (3) the therapeutic relationship is mediated at least to some extent nonverbally (for example, through the therapist mirroring the patient's movement); (4) movement contains a symbolic function and as such can be evidence of unconscious process; (5) movement improvisation allows the patient to experiment with new ways of being; (6) dance movement therapy allows for the recapitulation of early "object relationships" (the earliest interpersonal context) by virtue of their largely nonverbal nature.

It can be seen here that the theoretical framework to which dance movement therapy most frequently refers is psychoanalytic or psychodynamic (cf., Bloom, 2006; Siegel, 1984). This is not so much the case with von Sivers' development of therapeutic eurythmy, which came out of Steiner's anthroposophy (Steiner & Usher, 2007), nor with Albert Pesso's development of the Pesso-Boyden system of psychomotor therapy (Pesso, 1969, 1990). Dance movement therapeutic practices are also frequently Jungian in their theoretical orientation. This is notably the case with the authentic movement therapy developed by Mary Starks Whitehouse in the 1950s, as a kind of "free association with the body" that would allow participants to experience "movement in depth" and thus to express themselves fully (Behar-Horenstein & Ganet-Sigel, 1999; Chodorow, 1991; Pallaro, 1999).

There are many other western bodywork modalities involving movement that have developed in the past few decades that could be described here (Halprin, 2008). These would include the following: Sweigard's ideokinetic methods (Bernard, Steinmuller & Stricker, 2006; Sweigard, 1988); Emilie Conrad's continuum dance methods (Conrad, 2007; Gintis, 2007; Mchose & Frank, 2006); Elaine Summers' methods of kinetic awareness, which developed from her studies with Gindler's students Selver (whose work was discussed in Chapter 5) and Speads, who developed a system of "physical re-education," (Saltonstall, 1988); Bonnie Bainbridge Cohen's methods of body-mind centering (Cohen, 2008; Hartley, 1995); and

Steve Paxton's contact improvisation, which is influenced by the Japanese art of aikido (Kaltenbrunner, 2003; Pallant, 2006).

Despite their diversity and their multifarious history, many of these western lineages of bodywork had, by the end of the twentieth century, converged on the dual understanding that movement is necessary for the facilitation of bodily awareness, and that such awareness is essential to emotional and spiritual healing. While this realization may seem unremarkable, its radicalism must be appreciated in the context of European and North American cultures that not only have ignored the wisdom of the body, and not been responsive to the voicing of our embodiment, but also have forcefully promoted our alienation from the experience of our embodiment.

It is in this philosophical and cultural or political context that, toward the end of the century, Hanna and others started to advocate the use of the term *somatics*, which would designate bodywork methods focusing on the cultivation of proprioceptive and kinesthetic *awareness* and bodily *appreciation* (Hanna, 1993, 2004). This is, of course, central to the story of the emergence of somatic psychology and bodymind therapy. However, what was largely missing in the western traditions of bodywork that would support such therapeutic innovations was a language for understanding that the body is not only a system of structures and functions, as described by allopathic medicine, but is also the conduit for subtle energies. Although this understanding was intimated by Freud and developed by Reich, it has notably been the influx of wisdom traditions from Asia that supported the realization of this dimension of bodymind healing practices.

10
The Influx of Asian Wisdom

The contemporary impact of Asian healing disciplines and spiritual practices on the western cultures can hardly be overestimated. From our point of view, this impact occurred in three major waves. Knowledge of Asian wisdom traditions started to accumulate in the nineteenth century. In North America, this accumulation was less pronounced than in Europe, because the history of imperialist colonization had fostered the investigation of Asian cultures on the part of European academics and governmental officials. Historians often take the visit of Ramakrishna's disciple Swami Vivekananda to the 1893 Parliament of Religions in Chicago (and his subsequent tour of the United States lecturing on Vedāntic and Yogic philosophy, as well as general practices of Hinduism and Buddhism) to be a significant benchmark in North American recognition of eastern spiritual traditions (Vivekananda, 1947). The Theosophical Society, the transcendentalist poets, and various groups of spiritual seekers, became interested in the ideas or practices of these leaders, as well as other teachers such as Swami Ram Tirath and Aurobindo Ghosh. Although the Bhagavad Gita had been translated into English as early as 1785, it was not widely read until the end of the nineteenth century. About this time, spiritual teachings from other parts of the orient, such as the works of K'ung Fu-tzu (Confucius), Lao-tzu, and Chuang-tzu, also began to appear in translation. This set the stage for the further reception of Yogic and Taoist practices.

The second major wave in which Asian healing practices and spiritual disciplines came to influence western cultures resulted from the Chinese annexation of Tibet in 1949. The consequent diaspora brought many talented teachers from the distinctively Tantric (or Vajrayāna) version of Mahāyāna Buddhism to reside in Europe and North America. This is of particular significance to the history of bodymind therapy, because the Tibetan tradition offers some incomparably rich teachings concerning the embodied movement of subtle energies and our access to them (Barratt, 2006; Feuerstein, 1998b). More or less concurrent with the Tibetan diaspora, Daisetz Teitaro Suzuki came to the United States in 1951 to lecture and to take up a professorship at Columbia University. Suzuki's impact not only in bringing awareness of the Chán (Zen or Dhyāna) version of Mahāyāna Buddhism to the western world, but also of disseminating his knowledge of the Bahá'í faith, of Jōdo Shinshū (Shin Buddhism), Taoism, Christian mysticism (particularly the writings of Meister Eckhart and Emanuel Swedenborg), and the work of the Kyoto school of philosophy, was extraordinary (e.g., Suzuki, 1996). Suzuki's direct impact on figures like Jung, Fromm, Watts, and Thomas Merton is well known, but his influence on the western world is far more extensive (Fields, 1992). Buddhism and the other Dharmic traditions contain a wealth of teachings about the nature of consciousness, and these have gradually begun to influence western psychology (e.g., Davidson & Harrington, 2001; Tsering & Rinpoche, 2006; Wallace, 2007).

What might be called the second wave of impact that Asian wisdom has had on European and North American culture segues almost seamlessly into the third, which involves the cultural changes of the 1960s and their aftermath. It is challenging to avoid either overestimating or underestimating the significance of this decade as a period of cultural change. What is certain is that, since the 1960s, interest in the diverse traditions of Asian wisdom, as well as the accessibility of these teachings to the public, has blossomed dramatically. What is also clear is that very often these teachings are distorted even as they are assimilated into western culture (cf., Said, 1979, 1994). On the one side, the western culture of the imperialist powers has traditionally treated the third-world as a foreign "other" to be derogated and dominated. On the other side, there is a contemporary tendency to idealize eastern cultures as the exotic "other" that is the exclusive repository of all the spiritual wisdom that the west perhaps appears to have lost (Barratt, 2006).

Traditionally the cultures of Europe and North America were almost exclusively endowed with the doctrines of the Abrahamic religions (Judaism, Christianity, and Islam), which have generally viewed the body as an impediment to spiritual realization (e.g., Brown, 1988). Of course, there are notable exceptions to this in the Kabbalist teachings of Judaism, the mystical and Gnostic teachings within Christianity, and the Sufi teachings within Islam (e.g., Bischoff, 1985; Kugle, 2007; Pearson, 2007). Since the 1960s, western cultures have increasingly assimilated teachings from the Dharmic and Taoic religions. The Dharmic religions include the four great South Asian heritages (Hinduism in all its varied forms, Jainism, Buddhism, and Sikhism), all of which involve Yogic teachings. The Taoic heritages include the East Asian religions (Taoism, Confucianism, and even Shinto). These diverse traditions have in common a quite different attitude toward embodiment (Coakley, 2000).

To document fully the impact that Asian healing disciplines and spiritual practices have had on western psychology and psychotherapy, as well as other ancillary disciplines, would be an oversize task, and there already have been several interesting and important efforts in this direction (e.g., Ajaya, 1983, 2008; Bankart, 1997; Bolen, 2005; Coster, 2003; Doi, 1971; Epstein, 1995; 2008; Mruk & Hartzell, 2007; Paranjpe, 1984; Piano, Olson, Mukherjee, Kamilar, Hagen & Hartsman, 2002; Rama, Ajaya & Ballentine, 1976; Safran, 2003; Sheikh & Sheikh, 1996; Suzuki, Fromm & DeMartino, 1960; Unno, 2006; Watts, 1961; Welwood, 1979, 2002; Zhang, 2006). Instead, we will focus here on the Dharmic and Taoic treatment of embodied experience, and specifically on the wisdom of appreciating our embodiment as a system — container and conduit — of subtle energies.

To begin with some sweeping generalizations, it might be said that, compared to many of the Asian traditions under consideration, the ideology of the west tends to uphold the supremacy of rationalism, action-oriented will, the autonomy of the individual within the social and cultural context, and faith in the ideals of natural-scientific or technological progress (Doi, 1971). As was discussed earlier, the modern era of western culture — the era that is now coming to a close — established the hegemony of a dualistic, reductive, positivist and instrumentalist program of science. This acculturation assumes the splitting of mind and body, thus fostering the alienation of our embodied experience. It need hardly be stated that this alienation is quite congenial to much of Abrahamic theology and cosmology. By contrast, Dharmic and Taoic philosophies generally tend to offer a more holistic view of the body and of the universe, focusing on health and spiritual growth as a process of deautomatization involving the cultivation of awareness — a holistic reawakening of the senses, as well as a confrontation with whatever obstructs the free-flow of spiritual and emotional energies. Let us examine this in the context of the Yogic tradition.

As has been well documented, the Yogic tradition is spectacularly multifaceted, often embracing conflicting or competing doctrines (Feuerstein, 1998a, 2000). From the standpoint of the rather distorted depiction of Yoga that is currently popular in Europe and North America, it perhaps seems paradoxical even to assert that Yoga is primarily a spiritual-emotional discipline. In the west, Yoga is now so frequently identified with the physical activities of Hatha-Yoga, the use of postures and movements as a sort of body-cosmetic calisthenics that the essential feature of all Yogic practices as the facilitation of awareness is often lost.

The attitude of Yoga practitioners towards the body can vary significantly. Yoga's history begins in the Vedas, several millennia before the Common Era, in which it is alluded to as a meditative practice. By the time of the Upanishads, it is clearly a methodology for working with consciousness. The "classical age" of Yoga is usually identified with Patanjali's Yoga-Sûtra, which was strongly influenced by the Sāmkhya philosophical tradition rather than the Vedāntic, and was written or compiled in the second century of the Common Era (Feuerstein, 1989; Vivekananda, 2007). In preclassical and postclassical Yoga, the intent of the practice is union with the transcendent or *purusha*, which has been lost by the individual personality or *jîva*. There are several different Vedantic interpretations as to why and how the individual self became alienated or estranged from the transcendent "Self" and several different prescriptions for overcoming this condition and achieving *samādhi* or enlightenment. The purpose of physical practices, or Hatha-Yoga, would therefore be to prepare the body for enlightenment, to drop egotistic consciousness, and even to immortalize our embodiment. However, as Feuerstein (1998a) indicates, these nondualistic positions do not exactly fit Patanjali's account, which assumes "Spirit" to be irreconcilable with matter or with the bodymind. Patanjali emphasizes the intent of Yogic practice to stop mental chaos, or flow, thus quieting the entire bodymind, in order to experience the "Seer" (the Witness-Consciousness, in more Buddhist language the "Compassionate Witness," or in Sufi terminology the "Beloved"). The purpose of physical practices in this context, which is called Rāja-Yoga, is to transcend the body and stand apart from it. However, as can be seen, working with the body is — in one way or another — always considered essential to spiritual progress.

These variations obviously open the way for diverse somatic practices, which could range from the renunciation of physical desires (for food, for sex, and so forth) to their disciplined indulgence. However, what remains true is that — perhaps with the exception of some of those lineages which are more or less wholly antagonistic toward the body — the centerpiece of Yogic spiritual practice is the cultivation of awareness, including bodily awareness, and spiritual liberation or enlightenment is presented holistically as a bodymind eventuality. The importance of this point is that all these different Yogic lineages offer us a distinctive understanding of the nature of our embodiment, and this is perhaps the most significant issue for their impact on the rise of bodymind therapies in the western world.

The Yogic understanding of our embodiment is that the body is not only a physical entity of anatomical structures and functions, as designated by allopathic medicine and western science (Kaminoff, 2007). It is also a body of subtle energies. This is the esoteric dimension of our existence, a subtle anatomy that is "in but not of" the ordinary body of organs, bones, flesh and blood. In many ways, this is the essential feature of Asian wisdom that shakes up western psychology — the realization that there is a system of psychospiritual energy, or *Shakti*, that is the lifeforce pervading our embodiment (Fields, 2001). Āyurvedic medicine, which is sometimes considered the healing branch of Yogic science, also contributes substantially to this realization (Frawley, 1997, 2000; Lad, 2001, 2007; Ninivaggi, 2008; Pole, 2006; Wujastyk, 2002). Along with this tradition, Tantric practice, which is a set of methods of spiritual energy work aimed at enlightenment and is sometimes viewed as the parent of Yogic discipline, also contributes powerfully to this realization (Barratt, 2006; Carlisi, 2007; Feuerstein, 1998b). Currently, western science is being challenged to acknowledge that our embodiment is not merely physical, it also has a "supra-physical" double, in that there is an "astral body" or subtle energy body which Yogic science has known about for millennia.

The energy of the subtle body — usually called prānā — typically moves along channels or conduits, called *nādîs*. It is often said that there are 72,000 of these channels, although claims to experience up to 300,000 have been made. There are several main nādîs. In āyurvedic theory, these are more likely to be associated with the conventional physical structures of blood vessels and nerve pathways, and there are thirteen of them. In this theory, there are also one hundred and seven sensitive zones — somewhat like the Chinese *meridians* — called *marmans*, which are located at the connections of physical structures, but which also are focal points for prānic energy, such that disease occurs if the flow of energy is blocked around these points. In Hatha-Yoga, the nādîs have little association with ostensible physical structures and at least fourteen main ones are counted. There is some variation in this; for example, the *Yoga-Upanishads* mention nineteen. In all these slightly differing systems, the three main nādîs are the circuits running up and down the spinal axis — the *sushumna, idā*, and *pingalā*. These are particularly important for the *kundalinî* flow of the life-force through these channels, which constitutes a powerful method of psychospiritual clearing and personal growth (Jung, 1999; Khalsa & O'Keeffe, 2002; Krishna, 1997; Saraswati, 2001; Shannahoff-Khalsa, 2007; Mookerjee, 1981).

As is well known, in Yogic and Tantric teachings, the flow of prānā has been mapped to show a series of organizational centers called *chakras* (Dale, 2009; Judith, 2004; Karagulla & Gelder Kunz, 1989; Myss, 1996). There are usually seven of these counted within the body (there is often additional reference to chakras outside the body). However, Tibetan Yoga counts only five (condensing the throat, third-eye, and crown chakras). Somewhat similarly, Sufi Yoga talks only of four regions of divine life energy within our embodiment (Ernst, 1997; Douglas-Klotz, 2005; Helminski, 2000; Khan, 2000; Kugle, 2007).

Typically, the chakras are addressed in ascending order. The *mūlādhāra* is the root chakra situated at the perineum, and is associated with passion. The *svādhishthāna* is located slightly higher in the pelvis, and is associated with flow or the ability to change. The *manipura* is situated at the navel, and is associated with power. The *anāhata* is located at the heart, and is associated with caring. The *vishuddha* is at the throat, and is associated with truthfulness or speaking-out. The *ājnā* is the third-eye, and is associated with insight or perspicacity. The *sahasrāra* is situated at the crown, and is associated with freedom of spirit or ecstasy. There are also chakras in the hands and feet (which is significant in terms of the comparable Chinese medical system). Each chakra has specific psychospiritual characteristics and can be the site of energy blockages which have psychological and physical consequences (e.g., Breaux, 1989; Crawford, 1990; Cross, 2006; Fields, 2001; Johari, 2000; Kasulis, Ames & Dissanayake, 1993; Rao, 1982; Tansley, 1984, 1985). It is evident that what is offered in these schemas is nothing less than a "technology" for bodymind therapy and for spiritual growth (although this term is not really appropriate, since what is entailed is quite different from any technology that implies an instrumentalist or dualistic relationship with its subject matter).

There are several other ways of considering the subtle energies of the human bodymind that have progressively infiltrated professional healing practices in the western world. It would be remiss if the special contributions of Tibetan spiritual practices and healthcare methods were not emphatically mentioned here — because within Tibetan Buddhism there are extraordinarily sophisticated methodologies for meditating with the body, and cultivating an acute awareness of all the spiritual-emotional movements of energy that occur within our embodiment (Ray, 2002a, 2002b, 2008). The unique and unconventional methods of Vajrayāna or Tantric Buddhism, which is a radical development of the Mahāyāna path (and comprises the third yāna, or turning-of-the-wheel, of Buddhist development) focuses on the use of our embodiment in spiritual-emotional practice (which will be discussed further in Chapter 16). Through the combined use of visualization and breathwork in the practice of meditation, Vajrayāna discovers modes of access to the three implicated dimensions of our embodiment: the *nirmanakāya*, which is the body of flesh and blood; the *sambhogkāya*, which approximately corresponds to our understanding of subtle energies; and the *dharma-ya*, which is the pure light body of our Buddhahood and is also called the "body of reality itself." The distinctive use of visualization and the ease with which Vajrayāna practices operate within the imaginal realm reflects how this tradition developed out of the encounter of Mahāyāna from India with the indigenous Bön shamanism that antedated it in Tibet. In this respect, the significant aspect of visualization is that it can bring together our awareness of intentionality with the movement of subtle energies within and around our embodiment. We will discuss this further in Chapter 11. Tibetan methods of spiritual-emotional healing and personal growth are now available to western practitioners, not only because of the popularity of the Dalai Lama (although his spiritual leadership has opened westerners to the practices of his tradition and of the *Geluk* lineage that he heads), but also because of the recent accessibility of teachings from the *Nyingma* and *Kagyü* lineages, including those of Chögyam Trungpa, the charismatic teacher of "Shambhala Buddhism" (Trungpa, 2001, 2004, 2007). This development has profound significance for the future of somatic psychology and the rise of bodymind therapies.

Other than the Yoga of South Asia, the Chinese doctrines about the passages of chi energy are perhaps the system that is best known in Europe and North America today, coming to the western world long before Tibetan practices became available (Kuriyama, 2002).

But there are also systems which draw both from the South Asian and the East Asian traditions. The best example of this is the *sen-sib* energy system associated with Thai massage, which has gained western popularity in recent years (Apfelbaum, 2003; Chow, 2004; Gold, 2006; Salguero, 2004). Sen-sib is developed historically as a system influenced by ancient teachings from both India and China, and is now a somewhat distinctive model of embodied pathways within which subtle energies move.

Legend has it that traditional Chinese medicine originated over two millennia before the Common Era, and has developed into a highly complex system based on careful and systematic observation of the energy forces in the human body, in the natural world, and in the cosmos (Kuriyama, 2002; Unschuld, 1988). Traditional Chinese medicine includes dietary practices and herbology (based on a theory of the five elements), as well as many other methods, including those of acupuncture and acupressure. For our purposes, the latter are of particular interest because they are based on tracking the flow of chi energy (in Japanese, *ki*) through the body's pathways (Lu, 2005).

Just as prānā (or chi) is, in an important sense, supra-physical, the embodied channels along which it flows do not have exact anatomical or histological correspondences (Holland, 2000). These channels are called *meridians* (in Japanese, *keiraku*). There are twelve standard meridians, which run from the hands and feet, along the arms and legs, and connect with various biological functions (such as those of the lungs, large intestine, stomach, heart, and so forth). There are also eight extraordinary meridians, which are associated with other functions. As with all these energy doctrines, the issue of health and disease (both emotional and physical) is determined by the relative free-flow or blockage of the lifeforce through these various channels (Lu, 2005). Access to the channels, for the purposes of relieving energy blockages, is provided by *acupoints*, of which over four hundred have been mapped. As is well known, acupoints may be stimulated by needles, by manual pressure, or by the application of heat. The significance of these healthcare practices for addressing the patient's emotional life is noteworthy, for it is holistic and radically divergent from the biomedical model that has dominated western psychology (Zhang, 2007).

There are many additional healthcare practices or derivatives associated with traditional Chinese medicine, each using the wisdom of chi energy. These include the *tui na* system of massage (in

Japanese *anma*) and the Japanese tradition of *shiatsu* bodywork, as well as the "internal martial arts" of *tai chi chuan* and *qigong* (Beresford-Cooke, 2003; Jahnke, 2002; Liang & Wu, 1996). Other martial arts and heal-ing practices based on a bodymind theory and associated with these traditions would include a diverse group of more recent innovations, such as Morihei Ueshiba's aikido (Ueshiba, 1999; Ueshiba & Ueshiba, 2008), Mikao Usui's reiki methods of healing (Miles, 2008; Stein, 1995; Usui & Petter, 1999), and westernized practices such as "jin shin jyutsu" and "jin shin do" (Burmeister, 1997; Teeguarden, 2002).

Everything that has been said about the yoga tradition in this chapter could be reiterated in terms of the Asian practices of medit-ation. With the third wave of influx of Asian spiritual and healing practices, many different methods of meditation have become avail-able to the western public. It is not within the scope of this volume to explore these, except to indicate that — much like other Yogic practices — some of these methods seem to be based on a repudiation of the body and an effort to overcome the influence of its energies. Yet other methods have become essential for the practices of listening to the voice of our embodiment. For example, "mindfulness meditation" (or *Vipassāna*), which is mostly derived from the Buddhist tradition, is integral to the practice of embodied listening, to following the energies of our embodiment, and to increasing our sensitivity to the meaning-fulness that is within our depths (e.g., Barratt, 2004a; Hanh, 1999, 2008; Kabat-Zinn, 2006, 2007).

It is readily evident that the arrival of all these doctrines and practices into western culture throughout the twentieth century and especially during and after the 1960s provides a rich context for the emergence of somatic psychology and bodymind therapy. It seems appropriate to end this chapter by illustrating the rich-ness of these traditions of Asian wisdom with three quotations, each of which suggests the extraordinary acuity and exquisite sens-itivity of embodied awareness that was achieved through Yogic practices — and that are available to all of us today through the influx of Dharmic and Taoic teachings into the western world. The first quotation is from the Chandogya Upanishad, which was written sometime during the eleventh to fifth centuries before the Common

As large and potent as the universe outside, even so large and potent is the universe within our being. Within each of us are heaven and earth, the sun and moon, lightning and all the myriad of stars. Everything in the macrocosm is in this our microcosm.

This wisdom is echoed in a verse from Lao-Tzu, who lived about the time of Gautama Buddha, approximately fifth or sixth century before the Common Era:

Without going out my door, I can know all things on earth.
Without looking out my window, I can know the ways of heaven.

And echoed again in the Tantric teaching of Saraha, a Tantric adept who lived around the eighth century Common Era, and is considered one of the great Hindi poets:

Within my body are all the sacred places of the world, and the most profound pilgrimage I can ever make is within my own body.

11

Shamanic Practices and Transpersonal Psychologies

If there is a sense in which the entire universe exists within the energetic composition of every human body, then there is a sense in which the subtle energies of our embodiment impact the entire universe. This is a central aspect of the agenda of transpersonal psychology; although more aptly, we will refer to them in the plural, as psychologies, since they are a diverse grouping of perspectives. It is also central to the most ancient of spiritual healing practices — namely shamanism. Western knowledge about shamanic practices has paradoxically escalated in recent decades. The paradox is that, just when so many indigenous cultures all over the world are under attack and their way of life threatened with extinction by the socioeconomic forces of globalization, the peoples of Europe and North America finally seem more ready to learn from their wisdom. It is in this context that the potential impact of shamanic practices and transpersonal psychologies on the contemporary emergence of somatic psychology and bodymind therapies must be assessed.

The definition of shamanism is quite controversial (Francfort & Hamayon, 2001), but it seems workable to understand shamanism as a set of methodologies involving the use of altered states of consciousness for healing and for spiritual-emotional growth (Walsh, 2007) — altered states which Eliade (2004) notably and perhaps misleadingly called "ecstatic." In this context, every known indigenous culture has shamanic practitioners. Although such practitioners are often involved in naturopathic or herbological medical practices, the hallmark of shamanism is the use of altered states and, because there are many ways of altering consciousness as Barušs (2003) and others have amply documented, it must be added that shamanic practice always involves the contact made by the shaman's consciousness with realities that are ordinarily hidden, often referred to as the supernatural or "spirit world" (Harner, 1990).

Shamanism has long been studied by anthropologists and subjected to scholarly scrutiny (e.g., Kehoe, 2000; Jakobsen, 1999; Leete & Firnhaber, 2004; Maddox, 2003; Vitebsky, 2001), and efforts have been made to document its history (e.g., Maddox & Keller, 2003; Narby & Huxley, 2001; Price, 2001). Although there are indigenous forms of shamanic practice in both Europe and North America (cf., Cowan, 1993; McNely, 1981; Powers, 1986), it has been argued that the impact of shamanic practices in the twentieth century derives both from interest in Jung's work (Ryan, 2002) and from the 1960s popularization of Carlos Castaneda's writings on the Yaqui way of knowledge (e.g., Castaneda, 2008), despite the fact that Castaneda's connection with authentic shamanism has been challenged (Walsh, 2007). The influential work of Michael Harner must also be mentioned here, since he not only brought a wealth of knowledge about shamanism to the public's attention, but has developed a contemporary practice that he calls "neo-shamanism" or "core shamanism" (Harner, 1973, 1990). Currently, there is an enormous body of literature on shamanic practices and many readily available sources (e.g., DuBois, 2009; Pratt, 2007; Stutley, 2002; Walter & Fridman, 2004; Webb, 2004, 2008). Much of this contemporary knowledge of shamanism comes from South America, particularly the Amazonian regions (e.g., Harner, 1984). However, there are also many available sources on shamanic practices that come from Africa (e.g., Hill & Kandemwa, 2007), many parts of Asia (e.g., Connor & Samuel, 2001; Kakar, 1982), as well as Oceania (e.g., Wesselman, 2004), and many other parts of the world (Webb, 2008). On this topic there is a complex interaction between genuine anthropological scholarship and the popular imagination (Znamenski, 2007). In this respect, it seems advisable to examine the available literature appreciatively but not uncritically, and it is challenging to discern the exact relevance of this topic for the emergence of somatic psychology.

The use of psychedelics in shamanic practice is widespread (Furst, 1990; Harner, 1973; Pinchbeck, 2003; Walsh & Grob, 2005). The use of hallucinogenic states may be instructive for its illumination of the experience of our embodiment. However, in order not to overwhelm the issues under discussion, we will pass over this aspect of shamanism and focus on the findings derived from those altered states of consciousness that are not induced pharmacologically. Central to this topic is the imaginal practice of journeying as a method of healing (Villoldo & Krippner, 1987; Wesselman, 2003). Journeying includes the methods of "soul retrieval" (Ingerman, 2006; Villoldo, 2005) and of healing the subtle energies of our embodiment, which is sometimes called the "luminous body" (Villoldo, 2000).

"Imaginal" refers here to realms of experience attained, for the most part, by practices of visualization (Romanyshyn, 2002; Watkins, 1986). In general, traditional research on the psychology of visualization has treated the topic in terms of the cognitive construction of hypothetical possibilities (e.g., Ahsen, 1993; Finke, Ward & Smith, 1996; Kosslyn, 1980; Samuels & Samuels, 1980), which overlaps somewhat with the psychological investigation of the representational world of fantasy as the depiction of matters that are, in a conventional sense, unreal (e.g., Adams, 2004; Hall, 2007). However, in this context, a different ontological and epistemological claim is advanced; visualization is claimed as a method by which to arrive at dimensions of *reality* that are otherwise unavailable to ordinary states of consciousness, but which may be depicted allegorically (cf., Achterberg, 2002; Achterberg, Dossey & Kolkmeier, 1994; Noel, 1999). Thus, the shamanic practice of journeying is held to transport the participant to these realms, of some other spatiotemporal order, with the intent to retrieve or to heal an alienated aspect of the patient's own embodied experience.

From the standpoint of somatic psychology and bodymind therapy, the crucial issue here is to what extent the visualizations involved in journeying or other shamanic practices actually chart the inner wisdom of our bodily energies. This may well be how the Incas discovered "rivers of light" within our embodiment that coincide with the Chinese meridians (Villoldo & Jendresen, 1994). The allegorical form in which these journeys are retrospectively depicted — and which obviously varies greatly between one culture's idiom and another — is, in this sense, neither here nor there. What is critical is the extent to which the methods of visualization and "soul retrieval" — and the emotional release that often accompanies these methods — actually conjure and track the flow of subtle energies, remedying their blockages in the course of this momentum.

There is a Yogic-Tantric aphorism which states that prānā goes where intentionality goes. Cast in the form of a narration about other or outer worlds, the shamanic journey may well undertake an odyssey of healing within the inner realm of our embodied experience. This would be one way of considering the strong claims that are made for the ability of shamanic practitioners to heal psychospiritual ailments (e.g., Mindell, 1993; Rogat, 1997; Villoldo & Krippner, 1987; Wesselman, 2004), and especially those conditions caused by early, preverbal wounding (Gagan, 1998).

The distinction between outer and inner begins to dissolve as we enter the discourse of the subtle energy body. Practitioners of the visualizations sometimes involved in Vajrayāna meditation explore the interiority of the body's subtle energy systems, and occasionally they subsequently narrate their findings in picturesque allegories. Shamanic practitioners frequently explore realms that are allegorically projected as narratives about exterior events, but the effects of their practices are surely on the inner world of embodied experience. For example, the way in which a traditional Cuna shaman's journeying can in fact induce childbirth (in the event of a protracted labor) by narrating the passage of spirits within the woman's vagina and uterus, illustrates well the power of such practices. Details of this remarkable example have been given elsewhere (Barratt, 1993; Lévi-Strauss, 1963).

The question of the boundary between inner and outer is germane to the entire field of transpersonal psychologies. In North America, these psychologies established a presence within the field, subsequent to the 1960s, as a protest against the individualistic model promoted by the humanistic "third force" in psychology. Influenced by Jungian investigations, transpersonal research was also conceived as a part of a movement against the monolithic and hegemonic notion of natural-scientific progress, and as part of a postconventional impetus, characterized by perspectivalism and an openness to the possibility of the supernatural (Grof, 2000; Tart, 1992; Scotton, Chinen & Battista, 1996; Walsh & Vaughan, 1993).

Much of what has been achieved in transpersonal psychology in recent decades often seems disconcertingly disembodied and abstract. Indeed, it has occasionally seemed to some commentators that the entire enterprise of transpersonal psychology is constructed as an avoidance of the grounding of our self and world in our embodied experience. True to the least auspicious tendencies of the Abrahamic tradition, several erudite volumes on transpersonal theory are notable for their omission of the body, sexuality, or matters of subtle energies. For example, the otherwise admirable collection of essays on alternative ways of knowing in Braud and Anderson's anthology, the subtitle of which is "honoring the human experience," offers only a brief and rather uninformative section on bodily wisdom, and makes no mention at all of sexuality or of subtle energies (Braud & Anderson, 1998). Ferrer (2002) makes a sophisticated argument against experientialism in transpersonal psychology, by which he suggests that transpersonal phenomena are insufficiently understood — or even fundamentally misunderstood — when this understanding is referred to the individual's inner experiences. Against this standpoint, other transpersonal theorists understand that our experiential embodiment comprises the way in which we understand and organize the world, including the world of transpersonal phenomena (e.g., Deikman, 1982). The epistemological issues facing the future of transpersonal psychologies are, in any event, considerable (Hart, Nelson & Puhakka, 2000).

In ancient spiritual teachings, the esoteric transitions of almost every religious tradition — such as the *Shiva-Samhita* and the *Yoga-Shikha-Upanishad* — the universe is known to be discoverable within the movements of the human body, and the subtle energies of the human body are known to be connected with every other event in the cosmos. This provocative wisdom, depicting a universe of complex interdependence, is sometimes lost in contemporary transcendental psychologies — the discipline that one might expect to nurture this wisdom. Somatic psychology, however, appreciates our experiential embodiment as the source of all we can know. As this field continues to develop, it will be challenged to define more clearly its connections with the horizons of consciousness explored by transpersonal psychologies.

12
The Advances of Neuroscience

The awesome progress that has been made in the neuroscientific disciplines during the past decade promises to deliver powerful support for the agenda of somatic psychology and bodymind therapy. It also begins to diminish the credibility of those traditions in psychology that have limited themselves to the investigation of the representational mind or to the study of consciousness as a specific mode of reflectivity, the operation of which is somehow specifically localized in the cerebrum (cerebral cortex, limbic system, and brain stem). Obviously, any sort of general review of the advances of neuroscience is way beyond the scope of this book. This chapter will merely point to the way in which contemporary research challenges the modern notion of consciousness (that is derived from Cartesian philosophy), compels us toward a holistic vision of the bodymind, and intimates the potential relevance of quantum thinking to the further emergence of somatic psychology.

It is well known that Descartes viewed the mind as a nonmaterial entity which, lacking spatial extension, interacted with the pineal gland. He believed this pea-size endocrine gland which produces melatonin, and which — unlike much of the rest of the brain — is not separated from the rest of the body by the blood-brain barrier, to be the "seat of the soul." In our contemporary context, Descartes' convictions may be viewed with some amusement. However, we might now wonder how long it will be before the prevalent notion that the exclusive seat of consciousness is somehow located within the cranium comes to seem equally silly.

It is frequently asserted that over 90% of what is known about the brain's functioning has been discovered in the past decade (e.g., Comer, 2009). Contemporary research goes far beyond what can be achieved by electroencephalography (which has been around since the late nineteenth century, but came into widespread use only in the mid-twentieth century). New noninvasive neuroimaging technologies are now available (such as computed tomography, positron emission tomography, single photon emission computed tomography, magnetic resonance imaging, functional magnetic resonance imaging, magnetoencephalography, and transcranial magnetic stimulation). These have all been developed since the 1970s, and offer unprecedented modes of access to the observation of brain functioning (Cabeza & Kingstone, 2006).

The availability of sophisticated methods of observing events inside the cranium has not allayed debate over the nature of consciousness (e.g., Northoff, 2004). For example, there are those who seem to believe that a complete neurophysiological explanation of reflective consciousness (and of the language competences on which reflective consciousness is held to depend) might eventually be forthcoming, and thus advance an ambitiously reductionist program (e.g., Crick, 1995; Koch, 2004). This strictly reductionist viewpoint has been argued vigorously by eminent theorists such as Dennett (1992, 1997, 2007). But, on the basis of an assessment of the neuroscientific evidence, other influential theorists continue to argue that — without returning to Cartesian dualism — *lived experience* must be acknowledged as more complex, and of a different order, than its neurophysiological substrate. For example, the Australian philosopher, David Chalmers, asserts that the way in which subjectivity arises out of matter is deeply mysterious, and that consciousness should be considered a dimension like time and space, which can only be explained by its own psychophysical laws (Chalmers, 1997, 2002; Chalmers, Manley & Wasserman, 2009). This viewpoint has led to the recent and yet more radical theories of "neurotheology" (D'Aquili & Newberg, 1999; Newberg, D'Aquili & Rause, 2002). Many knowledgeable theorists are adopting a "wait-and-see" attitude toward this debate; an example of this would be the extensive and very illuminating writings on the topic by Edelman (Edelman, 1990, 1993, 2005, 2007; Edelman & Tononi, 2001).

What is more conclusive and of great relevance to the mandate of somatic psychology is that our neurological system has great plasticity or adaptive flexibility, and that it does not operate in the manner of any artificially constructed machine — in this sense, the methods of artificial intelligence, computer simulation or computational modeling that were so intriguing in the 1960s and 1970s have lost some of their sway over the scientific community (Dreyfus, 1972, 1992; Dreyfus & Dreyfus, 1992). This is why in all the social, cognitive, affective and neurobiological sciences there is a recent acceleration of interest in the grounding of our embodiment (e.g., Semin & Smith, 2008).

What has also been subverted by the advances of neuroscience is the notion that the agential "I" is the hallmark of consciousness, and that this "I" necessarily knows what it is thinking (Llinas, 2001). This is one aspect of the convergence of Buddhism and contemporary neuroscience (Nauriyal, Drummond & Lal, 2006; Wallace, 2007). Undermining the prerogatives of the "I," the concomitants of conscious experience now appear as a maelstrom of observable events distributed across the entire neurological system — and, indeed, across the entire bodymind. Even those who maintain a strongly reductionist viewpoint, arguing that the operations of consciousness are fully reducible to biochemical mechanisms, now construct their theories in terms of complex neural networks ("neurons that fire together, wire together") and recognize that there are gradations and multiple modalities of consciousness (Koch, 2004).

Research on mirror neurons — neural networks that can activate spontaneously in response to the observation of a highly specific relationship between the subject and an "other" — is just one example of a pathbreaking discovery that illustrates the complexity of what we call consciousness, and further dispels the myth of the independence of individual development (Ramachandran, 2003, 2005).

There are now strong reasons to understand consciousness as extending beyond the cerebrum and indeed beyond the central nervous system. As Aposhyan (2004) points out, the conventional idea that consciousness emerges from the summated activity of the cerebral cortex may be seriously limiting and thus distorting our understanding of the human condition (cf., Kandel, Schwartz & Jessel, 2000; Purves, 2007). The peripheral nervous system is certainly found to have wisdom of its own. For example, the complexity of the enteric nervous system, which is embedded in the lining of the gastrointestinal organs, has caused it to be called a "second brain" (Furness, 2006; Standring, 2008). It is capable of somewhat autonomous operation, and has been shown to function even when its line of communication to the central nervous system, via the vagus nerve, is severed. Additionally, the somatic nervous system, which controls all nonreflexive muscular movement as well as processing the reception of all external stimuli (from touching, seeing, hearing), might be said to have a "mind of its own," which is especially relevant to the way in which trauma is recorded (cf., Blakeslee & Blakeslee, 2007; Levine, 2008; Rothschild, 2000). A further example of this point is provided by Porges' polyvagal theory of autonomic regulation, which suggests that it is not only the central nervous system that is capable of social engagement (Carter, Ahnert, Grossmann, Hrdy, Lamb, Porges & Sachser, 2006). It is found that the dual branches of the vagus nerve (which runs from their ventral root to the brain stem) control such bonding or engaging behaviors as spontaneous facial expressions, listening and vocalization; their connectivity impacts heart and respiratory rate.

The issue of vascular communication must also be mentioned here, since the fluid channels of the body not only convey information throughout the body but also interact with it. The vascular systems thus monitor, and are also affected by, the fluids they transport (Aposhyan 2004; Margulis & Sagan, 1986). The issue of cellular wisdom is both generally accepted and yet more controversial, although its relevance to our emotional life is under debate (Greenfield, 2001). The astounding amount of information contained within each cell of the body is well known. It has also been calculated that only 2% of our body's intelligence occurs across synaptic connections, while the rest occurs at the interface of cellular membranes. However, Pert's claims regarding the cellular memory of emotions, and the idea that "the mind exists in every cell of the body," remain to be substantiated by widely accepted research and many neuroscientists dismiss it as fanciful speculation (Pert, 1997). Given the complexity of biochemical communication that occurs across the trillions of cellular membranes throughout our embodiment, the search for the cellular basis of memory is now exciting attention at the cutting-edge of natural science (Allport, 2001).

It might be noted here that what are sometimes called connective tissue memories are also treated with skepticism by many mainstream scientists. Yet almost every sensitive bodyworker knows well the phenomenon whereby, when a particular muscular group is stroked or palpated, an emotional release will occur, often with an involuntary visualization that is likely some distorted version of a remembered stress or trauma. This is also evident internally, as experienced colonic therapists often refer to the colon as having "emotional musculature" and know well how emotional releases of a specific kind can occur as a result of their irrigation procedures. One striking aspect of this phenomenon is that the involuntary memory arising when a particular muscular group is addressed is often "new," in the sense that it was not previously available for recollection and not recognized as a memory by the subject's reflective consciousness. It seems probable that "involuntary memories" of this sort may be encoded in a nonverbal form in any and all of the body's connective tissues, including every part of the musculature, the fascia tissues, and even the circulatory system. However, this phenomenon has yet to be investigated systematically. Indeed the prospect of systematic research on the topic is quite challenging methodologically. However, investigations that are relevant to this topic are increasingly being undertaken.

More extensively investigated are various aspects of our psychoimmunological functioning that may be related to the phenomenon of connective tissue memories (Daruna, 2004; Wilce, 2003). One example of this would be the studies that show increased blood pressure when emotionally charged material is suppressed or repressed from consciousness (Wilce, 2003). It is well known that endocrine hormones are crucial to the mobilization of energy resources and to the growth of new life throughout the entire bodymind. What is increasingly discovered by systematic research is the complex modulated interaction between neural signals and our immune cells via the endocrine system. This accounts for many of the clinical phenomena in which mental attitudes affect the body's health (Harrington, 2009). It has profound implications for our understanding of the bodymind's functioning. For example, it implies that stress or trauma — whether physical or emotional — not only taxes or overwhelms our cognitive abilities to process the events, but impacts the neuronal system, the endocrinological system, the immune system, and thus the entire holistic functioning of our embodiment (cf., Courtois & Ford, 2009; Cozolino, 2002, 2006; Levine & Frederick, 1997; Rothschild, 2000, 2003; Van der Kolk, 1987, 1994; Van der Kolk, McFarlane & Weisaeth, 1996). In this connection, the work of Schore on the involvement of multiple psychobiological systems in the tasks of affect regulation is highly important (Schore, 1999, 2003).

These researches imply the inadequacy of a reductionist approach to understanding our bodymind, and signify the holistic impact that events within the body's innumerable microenvironments can have on our entire being. The conclusion that there is a particular kind of wisdom inherent throughout our embodiment which is encoded, for example, in the ancient structures of collagen, nerve fiber (including the glia cells), and cerebrospinal fluid, is not avoidable.

There is an increasing body of neuroscientific research on "how the body shapes the mind" — which means how other aspects of bodily functioning condition the operations of the central nervous system in general, and the cerebral cortex in particular. This is one of the most exciting directions of contemporary neuroscience, admirably reviewed by Gallagher (2005). It is research that certainly offsets the Cartesian imagery of a cerebral mind that governs the bodily machine. One eminent example of such investigations would be Damasio's research and his important "somatic marker theory," which suggests the extent to which cortical functioning actually depends on the body's "gut feelings" in the operations of reasoning (Damasio, 2000, 2003, 2005). The force of all these recent advances in neuroscience has rendered reductionist theories obsolete. These approaches isolated bodily systems or their components and attempted to detail their operation independently. Such theories have proved limited in their ability to account for the way in which bodily systems actually operate in their natural context. Rather, contemporary neuroscientific findings compel us to acknowledge the fundamental unity of our bodymind.

The advances of neuroscience, along with the displacement of the agential "I" and the dispersal of what might be called "consciousness" through the entirety of the bodymind, compel a reconsideration of the way in which the term "consciousness" is used (Johnson, 1987; Lakoff & Johnson, 1980, 1999). Some neuroscientists distinguish primary and secondary consciousness. The latter is what we have been calling reflective consciousness (for example, it is not the experience of the redness of some raspberries *per se*, but rather the "I" that can have a thought such as "Here I am experiencing and thinking about the redness of these raspberries"). Secondary consciousness produces Descartes' formula, "I think therefore I am." It is widely held — for instance by Lacanian theorists and by a sizeable grouping of philosophers since Charles Sanders Pierce and Ferdinand de Saussure — that the egotistic consciousness involved in this sort of secondary reflection is a product of our induction into language (Barratt, 1984, 1993; Lacan, 1972, 1977). Symbolic language — and there is a profound sense in which the "I" is a symbol and not the ground of our being — permits this sort of second-order operation. By contrast, primary consciousness is often called *awareness* in somatic psychology, and refers to a level of sensitivity and responsiveness to qualities or events that may be not even be externally observable (including emotional processes) but that cannot necessarily be translated into words (or that can be translated into words but with some loss of quality). Our friends in the canine and feline world clearly exhibit awareness, but it seems very unlikely that they engage in thinking about their ability to think.

Thus far, we have discussed the advances of neuroscience in terms of what is known about our ostensible embodiment — the operations of tangible anatomical and physiological systems — and we have sidestepped the topic of our embodiment as a conduit for subtle energy systems, such as were discussed in Chapters 10 and 11. This brings us to consider the question of quantum realities.

It has already been mentioned how complexity theory — nonlinear interdependent dynamics — implies that consciousness is an emergent property, more complex than the sum of its parts, and able to affect the systems that support it (Clayton, 2006). This has opened the way for an array of different theorists to recruit ideas from the so-called "new sciences" of quantum mechanics and astrophysics — which address the micro-level and the macro-level of reality — in order to demonstrate how higher-order consciousness might exist as something more than the outcome of neurophysiological mechanisms (Kafatos & Nadeau, 1990; Mindell, 2000; Valle & Eckartsberg, 1981; Wilber, 1985; Zohar, 1990). The strategy could equally be applied to consciousness in general, to what I am calling awareness, and to the issue of subtle energies. It has obvious cosmic and theological implications (D'Aquili & Newberg, 1999; Murphy, 2006; Newberg, D'Aquili & Rause, 2002).

As is now well known, the quantum is an indivisible entity of a quantity that is related to the energy and the momentum of the elementary particles of all matter. Quantum "mechanics" (which is not at all mechanical in the traditional sense) is the most fundamental framework for understanding natural events at an infinitesimal micro-level. One aspect of what is both disturbing and exciting about the new sciences is that the closer scientists scrutinize the reality of matter the more it appears to consist of nonmaterial information — as pure potentialities of matter or as pure potentialities of energy, but not quite either (Greene, 2004). One consequence of this is that there is now a strong and serious argument that consciousness might emerge on the level of quantum reality, rather than from the gross operations of cell assemblies, neural networks, and the like (cf., Hameroff, Kaszniak & Chalmers, 1999; Hameroff, Kaszniak & Scott, 1998; Kauffman, 2008). At the extreme, some theorists now argue that consciousness might be nonlocal, operating entirely without embodiment, and are suggesting that quantum reality supports the notion of extra-sensory modalities of information transmission, and thus for the possibility of what are sometimes called paranormal phenomena (e.g., Radin, 1997, 2006; Targ, 2004).

At a less disembodied level, the findings of quantum science certainly legitimate the possibility of the body and its surrounding universe being imbued with subtle energy movements that are not firmly anchored to observable anatomical and physiological structures and functions (Davidson, 2004). While there is serious interest in the parallels between quantum reality and the findings of mystical insight — an interest somewhat shared by such great scientists as Erwin Schrödinger, Werner Heisenberg, Wolfgang Pauli, Niels Bohr and Eugene Wigner — there is also a danger of an overextended "new age" riff on the topic. Even a distinguished contributor such as Chopra (1989) has been unkindly criticized for advancing arguments merely based on reasoning by analogy. The claims of bodymind therapies with titles such as "quantum touch" and "quantum energetics" are likely to be met, at best, with sympathetic skepticism by the scientific community; at worst, such branding is likely to be dismissed as a shoddy marketing gimmick. So it is important not to overextend this mode of legitimizing the important notion of subtle energy systems.

Nevertheless, what the advent of quantum mechanics has achieved for somatic psychology and the bodymind therapies is a new openness to the fact that there is much that the natural sciences have yet to learn about energy systems. The vindication of ancient doctrines of subtle energy, such as prānā and chi may indeed be imminent. While some scientists complain that quantum mechanics is being hijacked for the purpose of legitimizing doctrines that are not yet subject to sufficient experimental proof, it is also definitely true that there are more reasons than ever to take seriously the constitution of our embodiment as a field and a conduit for subtle energies. In this context, the work of Henry Stapp (2007, 2009) seems particularly promising.

The new sciences have amply demonstrated the complicity of the observer and the observed. This complicity raises important questions about the ability of consciousness to influence the physical reality of the brain and of other matters. Stapp shows how the physical and the mental are emphatically conditioned by each other and he seems to offer a different approach to the issues of knowing and being. Opposed to the conclusions of Penrose (2007), Stapp's explication of the role and nature of consciousness in the universe reconciles the deterministic aspect of events, as constructed in mathematical models of natural evolution (Schrödinger's equation) and the empirical aspect of human experience. The former constitutes the "rock-like" properties of matter, while the quantum collapse of the wave functions constitutes the fluidity of our experiential or "mind-like" properties. This notion of consciousness as involving wave-function collapse, which is often interpreted in terms of quantum decoherence, is not an issue which we need to discuss further here, except to point to its general interest as a backdrop for any theory of the awareness of subtle energies.

The future of somatic psychology and bodymind therapies lies not with Newtonian models of the way in which bodymind methods might modify anatomical structures and physiological systems for the benefit of health, nor with Cartesian models of the way in which the mind, housed somehow in the cerebrum, dictates and regulates the activity of the bodily machine. Rather, the future will involve further understanding of the bodymind as a holistic system, with the awareness of its energies — which is itself the energy of awareness — pervading the entire system. As ancient sages insisted, awareness — Witness-Consciousness, the Compassionate Witness, the Beloved or the Seer — is itself pure prānā (Barratt, 2004a, 2006). It may flow freely, or it may be obstructed by the machinations of representational consciousness along with the reflectivity of the "I" — and this is the critical issue for healing and personal growth. The entirety of our experiential embodiment, with its miraculous capacity for awareness, is indeed what Descartes would have called the "seat of the soul."

Section III
Current Challenges: Possible Futures

There are many who would say that the major impediment to the continued rise of somatic psychology is the failure of some commentators to understand that this discipline is a psychology *of* our experience of embodiment. Such commentators thus fail to grasp the radical difference between this approach and those disciplines that are merely *about* the body (psychosomatic medicine, "mind-body medicine," rehabilitation medicine, sports psychology, human factors engineering, and so forth).

There are, however, several other major impediments to the realization of somatic psychology's centrality to the fields of human science. For example, the forcefulness with which cognitive behavioral theories deliver a technology of manipulation usable to dominant social groups pushes against the movement to return to the investigation of experience. And the ongoing allure of psychoanalytic and psychodynamic practices delivering bogus — as well as authentic — wisdom also tends to obstruct the rising momentum of somatic psychology. In this section, I will address — directly or indirectly — some of these issues.

Correspondingly, there are many who would say that the contemporary impediment to the continued success of bodymind therapies is the lack of solid data justifying them as an effective and evidence-based mode of treatment. The editors of the distinguished *Handbuch der Körperpsychotherapie* (Marlock & Weiss, 2006) call for such a program of research, seeing it as necessary for the assimilation of bodymind methods into the mainstream of behavioral medicine and psychotherapeutic practice. However, I disagree that the lack of evidence-based studies of bodymind therapies is such a challenge to their future. Indeed, as I suggested in Chapter 8, the application of positivist standards to validate bodymind methods of healing may be seriously self-defeating and inherently flawed. There are two embedded issues here. The first concerns the problems of a positivist-empiricist definition of knowledge, and the second concerns the criteria of effectiveness.

In terms of the first concern, the limitations of positivist epistemology, particularly as applied to the entire field of psychology, have been extensively discussed elsewhere (e.g., Barratt, 1984). To imagine that the normative routines of the natural-sciences can be imported wholesale to the study of *psyche*, without substantial loss and distortion, is in error. The challenge is that the study of psyche requires a focus on what Bataille (1988) called "inner experience" which is not readily available for public scrutiny, intersubjective verification, and so forth (which is not to imply that inner experience cannot be studied with rigor and responsibility, as in the phenomenological program). Ferrer (2002), in a critique of what he calls the "empiricist colonization of spirituality," offers a particularly powerful set of arguments against the application of empiricist language, methods and standards of validation to the field of transpersonal psychology.

Ferrer challenges the notion of "inner empiricism," which has held sway over many of those who argue for the differentiation of the human from the natural sciences (cf., Washburn, 1994, 1995). He criticizes the neo-Kantian aspects of Jung's experiential epistemology as well as the idea — shared by writers as diverse as Jung and Maslow — that some psychological states are epistemologically self-validating (Hart, Nelson & Puhakka, 2000; Jung, 1968b; Maslow, 1970; Meckel & Moore, 1992). He also cautions us against some of the validational claims made for Vedantic and Buddhist doctrines (Hayward, 1987; Paranjpe, 1984). Most compellingly, Ferrer deconstructs Tart's claims about the consensual validation of internal phenomena (Tart, 1983), and Wilber's very influential and sophisticated agenda for a "broad empiricist" approach to such phenomena based on a principle of falsifiability (Wilber, 1990, 1998). The latter is particularly important since Wilber convincingly argues that the epistemic status of spirituality and its interiority needs to be reconsidered and reassessed by those who — in the name of "science" — are somehow closed to this dimension of human experience. Nonetheless, Ferrer shows that some of the specifics of Wilber's program (particularly his attachment to the Popperian criterion of falsifiability) are themselves overly narrow and would thus preclude not only some of the findings of transpersonal psychology, but also of its somatic counterpart as the truthfulness discovered in the course of bodymind therapies.

The issue of the truthfulness of our embodied experience will be mentioned again in Chapter 17. Here I merely want to re-emphasize how the truthfulness of our embodiment is *not* equivalent to the adaptation of our behaviors to prevailing social, economic, cultural, and political conditions. As suggested in Chapter 8, this challenges the entire platform of evidence-based treatment. The "evidence" to be evaluated is invariably that of the *effectiveness* of the treatment with respect to the individual's adaptation or adjustment to the prevailing social order. In so far as it measures adjustment to existing cultural, political and socioeconomic conditions, the criterion of effectiveness thus serves an inherently ideological function. Effective treatments perpetuate the dominant social order. Their impact on the truthfulness of the individual's potential for self-realization is at best undetermined, at worst malign. As stated earlier, the argument that evidence-based interventions adhere to the fundamental value of inner experience, or that their operation is an authentic healing of the psyche, that is directed at the liberation of the individual's potential, cannot be sustained.

The major impediment to the emergence of somatic psychology and the rise of bodymind therapies is not the paucity of evidence-based research in the area. Rather, it is the failure on the part of the advocates of this discipline to recognize and embrace its inherent radicalism.

While it is certainly true that the discipline of psychology's relationship with the human body has come along way since Sheldon's anthropometric efforts to correlate temperament and somatotypes, the progress currently being undertaken in this field would almost certainly have been unimaginable to Sheldon himself (cf., Sheldon, 1940). However, for somatic psychology to realize its radical potential will require yet greater feats of creative imagination, and it will require the discipline to face some of its most fundamental philosophical and practical challenges.

In this third section, in the spirit of radical thinking, this section will offer a series of brief — and hopefully provocative — essays on what I consider to be the major challenges faced by somatic psychology. These are challenges that I believe must be met if the destiny of somatic psychology is to be secured. Chapter 13 will examine the delineation of the body, and this will include a discussion of boundaries and of the difficult questions surrounding the healing practices of touch. Chapter 14 elaborates some aspects of this discussion by addressing the inherently sexual dimension of our embodiment, as well as the challenges are thereby raised; this chapter also argues for the need for scholarship in a sub-discipline that I am calling *somatic sexology*.

Chapter 15 investigates some aspects of the political implications and ramifications of somatic psychology, and Chapter 16 discusses some of the spiritual impact of this science of embodied experience. Finally, Chapter 17 offers some notes on the possible future of human awareness.

13
Bodies and Boundaries

The issue of boundaries, and the question what constitutes a boundary, is essential to the entire discipline of psychology — although this centrality is not often acknowledged (cf., Akhtar, 2006; Lifton, 1976). Every inquiry within the human sciences explores and utilizes ideas about boundaries, even if it is not explicitly recognized that such notions are foundationally operative (cf., Wilson, 2004). In this chapter, some aspects of this notion are briefly reviewed, with a critique of unexamined "boundary-talk." This is followed by a discussion of the way in which the boundaries of our body are defined — including a commentary on the question where our embodied experience begins and ends. Finally, the boundary between the therapist and the patient is investigated, in terms of its ethical connectivity, and the controversial issue of the use of touch and other physical interventions for psychotherapeutic purposes will be explored.

There can be no question that we need boundaries, but it seems unlikely that we always need the boundaries that are given to us, or in the way that they are given to us. Many boundaries seem inevitable. Their construction is inherent in whatever language or representational system within which we think, and hence they appear to us as an unquestionable reality. The mug sitting on my table is a boundary-based perception, such that I am confident that there is "reality" to the table and the mug (and to the boundary where one begins and the other ends), as I apply these differentiated concepts to their perceptions. The boundary between most mundane physical objects and whatever is not that object seems quite unshakeable to ordinary states of consciousness.

However, other boundaries are conspicuously matters of social convention — even if the convention holds great power over our behavior. Consider the force of the traditional boundary between single and married sexual life. There is a significant list of things one might be able to do as single but not as married, and vice versa. Some social boundaries are more like permeable membranes — boundaries that can be crossed cautiously, with passageways that allow beliefs and behaviors from the other side to be explored intermittently or experimentally. An example of this would be the orgiastic May Day celebrations in some parts of medieval Europe, which allowed a temporary relaxation of sexual mores with respect to the expectation of marital exclusivity. In general, *boundaries are the codes that constitute our culture* and they are often culturally specific. For instance, a burp at the conclusion of a meal is a required signal of appreciation in one setting, but is considered disconcertingly rude in another. Some socially constructed boundaries seem extremely rigid with draconian consequences for those who violate them; an example would be the death sentence for adulterers, especially female ones, that is practiced even today in some parts of the world.

Often the existence of a boundary seems absolutely necessary, but where the actual boundary is positioned seems more arbitrary. As an example, the institution of legal marriage is rarely called into question, although it is arbitrary in the sense that one can imagine a society operating without it. However — as has just been suggested — individuals and cultures vary greatly as to whether sexual exclusivity should be part of the marital contract, whether premarital life should abstain from partnered sex, and so forth.

Even more deeply encoded in the fabric of our social arrangements and our personal attitudes is the necessary operation of an incest taboo — operating without such a taboo, if not unthinkable, would certainly be madness (cf., Shepher, 1983; Stein, 1984; Turner & Maryanski, 2005). Yet what constitutes incest is somewhat malleable. Procreative sex with a first cousin is condoned in some cultures, forbidden in others. Often the need for a boundary seems obvious, but its positioning seems somewhat arbitrary.

Although the boundary between physical objects does not appear to us as our creation, most sociocultural boundaries — including membranous ones — are clearly performative, in that they require our observance for their maintenance, and sometimes we do not even know they are operative until we transgress them. In this regard, almost every individual has recollections of such experiences: childhood memories of being innocently seductive toward a parent, and then shamed or scolded for it; minor episodes in adulthood when one arrives at a function in casual dress only to discover, with some embarrassment, that formal attire was required; unwittingly doing the wrong thing, in the wrong place, at the wrong time. Conversations in psychotherapy and counseling frequently recount such events.

Social boundaries can be perpetuated both in their condoned observance and in their condemned transgression. Again, think here of how arbitrary and malleable the boundary between "married" and "unmarried" life may be, why crossing the boundary is performed so repeatedly, and why the distinction between appropriate conduct on the one side and appropriate conduct on the other side is — in almost all known cultures — so frequently and fervently debated. Think here of the frequency of salacious gossip, or the repetitive themes of sitcoms: "Was it okay that he kiss his friend's wife on the cheek, or was he going too far?" ... "Is her short skirt alright to wear around the house, but too suggestive to be worn in public?" ... "Is a consensually open marriage *wrong* under any circumstances?"

The operation and maintenance of these sorts of boundary requires their repeated enunciation and usage — they are, after all, matters of social convention. They may be deeply embedded in our sociocultural codes (and less likely to change with history), such as the boundary between "good" and "evil," or they may be less deeply embedded and more mutable, such as the boundary between "polite" and "impolite" behavior. We probably all sense a difference between the fundamental lawfulness of some boundaries (such as the necessity of the incest taboo) and the normativity of cultural codes that are more clearly arbitrary (such as the prohibition on continuing to drive when a red traffic light is illuminated). But this sense varies across cultures and historical epochs. This is why anthropologists and other cross-cultural researchers, as well as historians, have so much to tell us about their operation.

Contemporary clinical practice is full of talk based on boundaries and about boundaries (e.g., Akhtar, 2006; Celenza, 2007). So too is our popular culture. Consider the lamentable frequency with which many clinicians invoke boundaries between what is "appropriate" and what is "inappropriate." Here is a veritable goldmine — or minefield — of boundaries which many clinicians and coaches all too readily believe they have the authority and the expertise to prescribe ("it is appropriate to do this, inappropriate to do that"); more rare is the therapist who will explore open-mindedly with patients the effect of the cultural mores and beliefs that have been impressed upon them. Most clinical professionals do not see themselves for what they are — peddlers of moralizing advice, acting in the manner traditionally characteristic of many clerics. Rather, if they reflect on their own activities at all, they typically view themselves as purveyors of a scientifically-based expertise as to the adaptivity of different behaviors within their social and situational context. Yet the effect of their activities is ideological, and often harmful. The invocation of *appropriateness* may serve to produce a more adjusted individual; it also serves to reproduce the social order, marginalizing and isolating those who do not conform sufficiently to the precepts and tenets of that order.

We might note here how ideologically powerful is this *notion of appropriateness*; not least because there is an intrinsic elasticity in the situational boundary to which it refers (the criteria of appropriateness are usually invoked by one who has power over the other for whom the invocation is supposed to pertain). Appropriateness does not have the acuity or fixedness associated with "right/wrong" or "true/false," and is all the more pernicious because of it. Appropriateness is the sort of boundary around which there are always people in authority, self-appointed or solicited, who are ready to tell us how to conduct ourselves. It is also the sort of boundary that we may not know is operative until we bump up against it or actually transgress it. In this way, the multiple boundaries of appropriateness serve to maintain and reproduce the social order of behavior. So too does that other dangerous boundary, the distinction between what is "normal" and what is "abnormal."

The invocation of this highly-charged and ideologically-steeped boundary is the benchmark of *clinical authority*. The clinician is professionally endowed with the authority to discriminate what is normal from what is not. This discrimination is allegedly based on a scientifically-generated expertise on the operation of the boundary. Here I do not wish to dismiss the value of clinical insight, but rather to point to the danger of its unreflective or unexamined engagement. All too readily, the clinician becomes the professional purveyor of social codes, whose task is to define and differentiate what is culturally conventional and acceptable from whatever is considered weird, crazy, and pathological. There is great danger in conjuring boundaries without subjecting them to open-minded and rigorous examination. *Clinical expertise carries the danger of authoritarianism and mindless compliance with social arrangements that may well be oppressive.*

Our culture, both professional and nonprofessional, seems to be riddled with unexamined boundary-talk. Much of it is confusingly multifaceted and contradictory. Consider just one example. On one side, today's public is often urged by professionals (from reputable clinicians to self-help charlatans and media pundits) to secure firmly the physical and emotional boundaries between ourselves and our children, ourselves and our lovers, ourselves and our friends. We are also enjoined to love each other, and even in some contexts to cross — at least some of the membranous — boundaries that divide us from these other individuals. On another side, the public is extensively exposed to somewhat "new age" rhetoric, suggesting that boundaries are inherently *the* problem in human relations. At a summer solstice celebration I attended, participants were urged "just for today, to let down all boundaries between us" (of course, if were this possible, it would actually be a nightmare). This is one of the contemporary imageries of authentic intimacy, a self-proclaimed state of being relieved of all social boundaries, which reverberates profoundly with what some Jungians have called our "fusional complex" (Schwartz-Salant, 2007). It contrasts markedly with an equally virulent plea for stronger boundaries between those who are vulnerable and those who would exploit or take advantage.

Unexamined boundary-talk is an ideological trap, even though philosophical scrutiny of this issue does not necessarily resolve it. For example, some philosophically adept commentators argue for the importance of boundaries in protecting us from acts of aggression. Others argue that boundaries are themselves aggressive acts. Perhaps the security of boundaries protects us from an "ungodly chaos," or perhaps it prevents us from the "godliness of unification." Theodor Adorno, who was one of the twentieth century's great philosophers, spent a significant portion of his career demonstrating the inherent violence of all syntheses, and thus challenging the entire heritage of enlightenment philosophy that invariably preaches or presumes the harmonizing virtues of synthetic functioning (Adorno, 1966). Contrary to this dissent, notions of synthesis, harmony, integration and unification have been the ideological watchwords of the modern era: *e pluribus unum* (cf., Taylor, 1987, 1993).

Since the universe actually is — as the new sciences keep telling us — an intricate and fluid concoction of vibrationalities, how we draw boundaries between one thing and another is the foundational question of psychology. Expressed differently, this is the question of identity, including the identity of the psyche, in a universe of nonlinear dynamic interdependence (Wilson, 2004). Let us examine further the issue of identity in relation to bodily matters. Several interrelated points need to be made here, and it may help to bring into focus the following three vignettes:

- A mother bathes her son, carefully washing his penis and scrotum, while the little boy squirms and giggles. The child has just turned two years.

- A second mother, also bathing her son, carefully washes his penis and scrotum, while the boy squirms and giggles. The boy is now seven years.
- A third mother bathes her son, carefully washing his penis and scrotum. Her son squirms and giggles. He is a college sophomore.

The son's life begins inside the body of the mother and inseparable from it. Outside the maternal body, the young child's body is still intimately enmeshed with the mother's. Yet typically, the son's life evolves to a point of maturity wherein bodily contact with the mother is strongly forbidden, except perhaps in very limited and circumscribed ways. Reading these vignettes, most of us smile indulgently at the first dyad's behaviors, become somewhat concerned about the "appropriateness" of the second ("why hasn't the boy learned to wash himself?"), and are appalled at the conduct of the third mother-son duo. Even if we are now told that the third son has cerebral palsy, our uneasiness is typically not entirely allayed (and clinicians will typically make dire predictions about the mental health of both participants in the third scenario). For later consideration, we might also note that, even if it were specified that the woman involved was not the boy's mother or anyone biologically related to him, at some point in the progressive scenario we still think of the interaction as a form of misuse or abuse.

As psychoanalytic wisdom has well informed us, the boundary between what is "me" and what is "not-me" is complex, always fragile, and somewhat fluid. In general, it is a hard-won accomplishment, occurring on many levels, and spanning the entire course of personal development. Even on the level of our cognitive representations, the question when "me" began (let alone when "me" will end) is exceedingly challenging. If your name is Sarah, consider the following. When did the person "Sarah" begin? Answer this question not in terms of what is objectively known about the physical entity named Sarah, but in terms of your sense of identity and the psychological level of representational life. It is easy to assert that sometime in toddlerhood, you as Sarah began to articulate a "Me-Sarah" identity in what Lacan would call the "Symbolic" register of language. However, it would be an error to conclude that this is where and when your identity began. Even before conception, you as Sarah existed — in the twinkle of your father's eye, in your mother's childhood play with dolls, and so forth. In this sense, there was a "you" existing in the world of representations well before your conception, let alone before you were birthed as a baby, and long before you yourself articulated your identity in language. In this sense, you were inducted into your representational identity, which existed before you were yourself able to articulate its representation (and long before you were able to articulate your experience of yourself). In part, this illustrates the power of what Lacanians call the "Imaginary" register that is responsible for the social formation of our "ego" (Grigg, 2008; Rabaté, 2003).

Even on the level of our representational life, this *me/not-me boundary*, its elasticity and permeability, causes us a lifetime of difficulties. Consider here the ubiquity of "defense mechanisms" in our everyday psychology (Cramer, 2006). Projection and introjection are prime examples of ways in which the boundary of the self is both protected and, at the same time, distorted away from the consensual standards of accuracy. Projection begins with the denial of a thought or feeling that the subject is actually having, followed by its attribution to some other person or persons — "I am not having this thought or feeling, they are" (cf., Young-Bruehl, 1998). Our ways of knowing what we are actually experiencing are continually being distorted in the interest of the egotistic consciousness of our representational system maintaining its own sense of propriety and equilibrium.

Psychoanalytic wisdom has also had much to contribute to our understanding of the way in which bodily experiences shape the formation of a me/not-me boundary. Consider here the significance of the challenging and fascinating experience of defecation from a toddler's viewpoint (cf., Erikson, 1995). If you put your head between your knees, you can see the fecal stick, which is clearly a protuberant part of "me" — just as an arm is a protuberant part of me, but with somewhat different proprioceptive and kinesthetic feedback. Then this "me" drops away suddenly, and gets flushed or otherwise discarded. Understood in this manner, it is no wonder that toilet training acquires such importance in a child's life, because it is not just a matter of gaining control over the anal sphincter, it implies a radical challenge to, and modification of, one's sense of identity.

The issues of bodily experience in the shaping of the me/not-me boundary are yet more complex. The defecation scenario just described involves a comparatively simple interaction between a child's — supposedly realistic — representations of his or her body and exterior eventualities. However, as some of the most brilliant psychoanalytic writings have shown, our embodiment is not only perceptually, proprioceptively and kinesthetically experienced, it is also experienced in fantasy. An example which is found frequently in psychoanalytic practice — but which is greeted with skepticism by many who have not yet benefited from psychoanalytic exploration — is our ability to experience the whole body as a phallus (Lewin, 1933; Reich, 1953).

Such fantasies not only impact our sense of self and our behavior powerfully, but actually modify the perceptual, proprioceptive and kinesthetic experience of the body. Even skeptics recognize that patients suffering severe eating problems such as anorexia nervosa actually see their body as being markedly different from an objective — consensual — description; they actually perceive and internally experience their body as fat, when it is not (Minuchin, Rosman & Baker, 2004).

It can be seen here how the question where our embodiment begins and ends is far more challenging than might be supposed. It is usually suggested that our embodied experience is confined to the limits of our proprioceptive and kinesthetic sensations, which are then overlaid by our perceptual and conceptual representations of the body. This is congruent with what we know about the importance of skin sensuality in establishing for us where we end and the rest of the world begins. This has been well discussed in the psychoanalytic writings of Didier Anzieu (1989, 1990, 1995), Esther Bick (Piontelli, 1986; Sandri, 1998), and others. Since the pioneering work of Montagu (1971), the significance of the skin for our healthy functioning has been generally acknowledged. Bodymind therapists especially know well how important the sensuality of what is sometimes called the "skin envelope" is for emotional and spiritual health as well as physical wellbeing (cf., Anzieu, 1989, 1990, 1995). However, if we consider the body as an energy field, then our notions of embodiment and of our embodied senses might extend beyond this envelope. The circulation of subtle energies within and around the body is not confined to any space within the skin surface. Many energy psychologists refer to *chakras* that are additional to those usually enumerated and that are outside the body (Dale, 2009). Energy anatomy is not tightly anchored to anatomical and physiological structures, so there is no reason to suppose that it is confined within the body itself. This added complexity implies that it is possible to touch another person's energy field — or to "invade their space" — without the physical structures of the body being tangibly contacted (Bowler, 2004).

The interpersonal context of body fantasies also adds complexity to the issue of the me/not-me boundary. The complexity is interpersonal in the sense that the processes we have just described entail the possibility of experiencing one's body as not one's own — a problematic confusion of boundaries. In the three vignettes of a mother bathing her son, one might well ask "who owns the boy's penis and scrotum?"

or more accurately "in whose possession is the boy likely to experience his penis and scrotum as being?" The various sensations of the genitals may be his, but there is still a troubling sense in which the mother is treating them as her possession — and the boy may experience them as such. This is well illustrated in the case of "Cindy" reported by Goethals and Klos (1976); contemplating her first heterosexual intercourse, Cindy refers to it as "the critical cut against my mother, giving away her prize, my virginity" (p. 35). It is absurdly simplistic — and an ideologically mischievous distortion — to hold that every individual simply owns their embodiment. Although the days of slavery may be over, many of us find that significant aspects of our embodiment are indeed psychologically committed to someone — or something — other than ourselves.

Elsewhere I have discussed the complex transgenerational dialectics by which the father's unconscious dynamics are marked on the body of the son (Barratt, 2009a). Using an analysis of myth as well as my clinical experience as a psychoanalyst, I argued that the repressed unconscious is structured around dynamics of "deathfulness" and "castratedness" (Barratt, 2004b). Subjectivity develops (in this case male subjectivity) around the patriarchal transmission of these dynamics from the father (or paternal figures) to the son, and this transmission is engraved in the son's embodiment (Barratt, 2009a). The commonest illustration of this is the circumcisional cut which Abraham inflicted on Isaac (and on his whole entourage of slaves and others), which seems to be part of a bargain with God by which he might secure his symbolic immortality. Among the many other famous examples of symbolically "castrated" sons, whose body bears the mark of their father's ambivalence toward them, would be Oedipus who survives his father's attempt to murder him yet remains club-footed, or the elephant-headed Ganesha who survives decapitation at the hands of his father. There are additional depths to this thesis, which cannot be detailed here. For our present purposes there are three aspects that are important.

First, the findings of psychoanalytic and anthropological inquiry are compelling in their suggestion that, amidst the confusions about boundary issues — the admixture of their necessary and arbitrary properties — the incest taboo is a boundary of special significance. Even when they are fully operative in their regular performance, *boundaries elude definition except in relation to other boundaries*. However, much as the relativity of the material universe can be referred to an abstract point of origination, psychic boundaries can be referred to a "point" of origination, which we might call the *boundary-imperative of incest taboo.* In this sense, *incest is the "boundary of boundaries"* — similar to Lacan's idea of there being a "law of laws." We do not really know why or how the incest taboo operates. After all, it is obeyed even by those with no knowledge of genetics, and our obedience to it occurs naturally, without the need for any awkward reasoning about the potential problems of inbreeding. But we do know that, in one form or another, it is one of the few conclusively universal features of the human condition.

Second, the boundary of incest is pervasively marked in all the reactions and responses of our embodiment. It is deeply encoded within us. The excitement, anxiety, and revulsion that accompany even the idea of intercourse with one's father and mother — or with one's son and daughter — comes to us as if automatically and naturally, inscribed in the deep structures of our representational language and in the discourse of our embodiment. The instructive myths of characters such as Abraham, Oedipus, and Ganesha, as well as all the legends of parent-daughter relationships, are dramatic only in the extremity of visible damage done to the body (loss of the prepuce, club-footedness and blinding, decapitation). In another sense, these dynamics are — usually in more subtle form — expressed *within* the experience of every human body.

Indeed, it is primarily the incest taboo that marks the boundary between representational consciousness and that which it dynamically represses from itself. This is a crucial point that has not been elaborated enough in the psychological and philosophical literatures. The psychologically foundational consequence of the incest taboo is that it marks the boundary of reflective consciousness, establishing what is known as the "repression barrier" (the distinction between the acceptable realm of conscious-preconscious thoughts and feelings, and the forbidden realm of the repressed). This is a major reason why some psychoanalysts — including myself — insist that we are all "castrated," and indeed that we are egotistically endangered when we fail to acknowledge this intrinsic insufficiency or inadequacy of our subjectivity (Barratt, 2004b, 2009a).

Third, if the incest taboo is the prototype or "boundary of boundaries" and the boundary-imperative of incest taboo founds the structuration of all human discourse (both as representational life and as our experience of embodiment), then its impact governs every interpersonal relation (of course, including that between therapist and patient). This raises a double conundrum for the healing relationship. On the one side, our examination of boundary functioning suggests the derivative character and constitution of human intentionality, meaning that boundaries create and define actors, rather than actors creating or defining boundaries. On the other side, the boundaries operative in the healing relationship are socially positioned in a manner that is somewhat arbitrary, and in this respect the intentionality of the therapist is crucial to the ethical security of the relationship and hence to its potential for healing.

This brings us to the controversial issue of touch in the healing relationship, and the ethicality that defines healing. As is well known, there are socially and legally mandated — moralistic — codes around this issue. Consider the professional distinctions between a physician, who is allowed to touch every part of the patient's body (but only under conditions of emotional dissociation), a massage therapist or bodyworker, who is allowed to touch almost all parts of the body (although the conditions of emotional engagement are usually ill defined), and a psychotherapist, who is allowed to touch no part of the patient's body, except perhaps for a formal handshake (but who is not going to facilitate healing unless he or she is capable of an intimate emotional engagement with the patient). These are the social, legal and moralistic codes that contextualize these professions. However, the codes have almost nothing to do with the ethical processes and amoral practices of healing the psyche. Healing is an ethical calling, but it is also amoral in the sense that it does not necessarily have regard for the plethora of social codes and boundaries. In this respect, we know that touch facilitates emotional and spiritual healing — not the objectivating touch of the medical practitioner, which has mechanical purposes, but the emotionally, sensually and energetically meaningful touch of the bodymind therapist. It is precisely this type of touch that society monitors prohibitively, and the prohibitions have little to do with the sacred task of healing.

An insightful psychoanalyst once suggested that relationships intended to heal the psyche are characterized by three qualities: *safety, freedom*, and *intimacy* (Limentani, 1989). To this list, one might well add the *ethicality of truthfulness*. Healing the psyche is an ethical and spiritual process, although its conduct is amoral (in that it is not constrained and conditioned by social and cultural rules and regulations, and is "neutral" with respect to their force). I would argue that, given the nature of the psyche, these qualities are necessary dimensions of any authentic healing process, and I would define them as follows:

- Safety implies the patient's protection from physical or psychological harm. Yet more profoundly it also means that the patient's psyche is protected from the travesty of incest.

- Freedom has three aspects. First, it means that the patient's potential for self expression is facilitated to the fullest extent feasible. Second, it means that the patient's "free-associative" expression — uninhibited discourse that may relinquish the normative and normalizing codes of propriety, of logical and rhetorical convention, and even perhaps the laws of semantics, pragmatics and syntax — is given full license. In this sense, healing requires that the rules and regulations of the social order be, at the very least, held in abeyance. Third, freedom means that the healing process is freely engaged, and I would argue here that there is no genuine healing of the psyche under conditions that are coercive or authoritarian.
- Intimacy not only implies something about a nonjudgmental openness toward matters of intentionality and desire, it also suggests something crucial about the intense emotional engagement of the therapist and the patient that is necessary for healing to occur. It suggests that the limits of healing may be the limits of the therapist and patient's willingness to surrender to the healing process.

With respect to the issue of therapeutic touching, it is a mistake to imagine that a rule of abstinence disallowing the physical engagement of the therapist and the patient somehow ensures that their interaction is benign and without violations of the patient's being. It has to be remembered that, for the patient, the unconscious connotations of the healing relationship are always incestuous — this has been extensively researched and proven by psychoanalytic treatment methods (Esman, 1990; Mann, 1999; Rosiello, 2000). This means that the quality of verbal interaction can be experienced as much a violation of the incest taboo as any physical engagement — perhaps short of intercourse. Just as the notion of freedom implies that a genuine healing process cannot be coerced or compelled, the notions of safety and intimacy imply that authentic healing cannot be coaxed or cajoled. The seductive therapist, the persuasive therapist, and the therapist who is engaged in the healing process in order to accrue narcissistic gratification (such as the gratification of power and admiration) probably does as much damage as the "therapist" whose abuses are more explicit.

Given the healing power of touch, its prohibition in psychotherapy — however well intended — is absurd. Given the integrated nature of the bodymind, such a prohibition could actually be harmful, in that it promotes a mode of therapy that might well perpetuate the patient's alienation from his or her embodied experience. This is the nightmare of badly conducted psychotherapy as a seemingly endless procedure of talk and more talk — the perpetual telling and retelling of the patient's stories, which fails to facilitate the patient's potential to live in the present. Healing requires that the healing process touches every aspect of the patient's being-in-the-world ("body, mind and spirit" as is nowadays often stated). In this respect, a therapist's failure to integrate touch within a healing practice risks supporting the perpetuation of a patient's alienation from his or her own embodied experience. It can well be argued that to take such a risk with a patient — and to circumscribe therapy as only a talking procedure — is to fall short of the ethical standards required for healing.

Bodymind therapists and psychodynamic psychotherapists (and the distinction is, of course, thoroughly outmoded, and will be transcended with the further emergence of somatic psychology) both need courageously to take a stand for the necessity of both emotional and physical touch in the practices of healing. Again, the prohibition against the healing power of touch is an absurdity. If it were feasible to monitor, what would make more sense would be a prohibition against any emotional or physical touch that is indulged for the therapist's gratification. In this context, it would seem that the critical factor in facilitating a healing process is the therapist's attitude, conscious and unconscious. In this regard, bodymind therapists can learn from — the better side of — the psychoanalytic tradition. While it is sadly true that many psychoanalysts fall short of these precepts, psychoanalytic principles have always suggested that a *therapeutic attitude* has three facets:

- The therapist works to refrain from judging the patient's being-in-the-world (his or her thoughts, feelings, fantasies, actions, and bodily phenomena), or trying to mold these phenomena into something different, such as an ideal form, but rather the therapist works to elucidate their meaningfulness for the patient and facilitates a process in which the patient is newly empowered to listen to the meaningfulness of his or her experience.
- The therapist works to establish the therapeutic relationship that the patient is able to experience as safe, free and intimate.
- The therapist attempts to place himself fully and consistently at the service of the patient's personal growth and in the service of the healing process, by working to abstain from narcissistic and other gratifications in the relationship, including the gratifications of authority, power, sensual pleasure, admiration, and so forth — with the exception of the explicit recognition that the therapist has to

make a reasonable, but not exorbitant, living from professional practice.

Understood in this manner, therapy is not only a sacred calling, but a profoundly ethical imperative. The issue of touch — emotional and physical — is not so much a matter of what or where is touched. Rather, it is a matter of why it is touched, and the ethicality with which the touching process is undertaken.

Similar to the best of psychoanalytic practices, bodymind therapy stands in contrast with procedures that are authoritarian, directive or manipulative. In the context of the unique insights provided by somatic psychology, it can now be stated that the latter procedures, in which the clinician knows better how to lead the patient's life than the patient does, can never be genuinely healing, and the clinician who uses the patient for his or her own gratifications derails the healing process. Of course, the clinician who incests a client or patient crosses the most essential of all boundaries and enacts a coercive abuse of power that is not only antithetical to the process of healing, it is also the most heinous ethical violation of the patient's being-in-the-world that could be indulged.

The ethical dimensions of safety, freedom and intimacy cannot possibly be pursued unless the inevitable seductiveness of therapeutic discourse is contained by the therapist's rigorous practice of foregoing all narcissistic gratifications in establishing a relationship with the patient that, nevertheless, must be profoundly emotionally engaging. The challenge of bodymind therapy is for the community of therapists to assert the freedom of touch as essential to the healing process, all the while insisting on standards of training that secure each therapist's ability to create relations that are safely intimate. It must be firmly emphasized that, at some point in the majority of healing processes, not to touch would be an ethical lapse that perpetuates the alienation of our embodied experience.

14
The Inherent Sexuality of Being Human

If the issues of healing touch and the ethicality of touching challenge the further development of somatic psychology and bodymind therapy, then the issue of sexuality — the inherent sexuality of our embodiment — is an even greater challenge. It can be anticipated that the discipline will curtail its own potential if its practitioners continue to insist that healing the bodymind is not an act of sexual healing. In this chapter, some attitudes toward sexuality are briefly reviewed, and how sexuality might best be defined and understood is then explored. Finally, a discussion of the definition of sexual health is offered, and the implication of these ideas for the future of psychological practice in general and specifically for the future of somatic approaches to healing the psyche is explored.

In *Sexual Health and Erotic Freedom* (Barratt, 2005), an argument was introduced about the *sexification* of North American culture, particularly that of the United States. The thesis also applies — although perhaps less conspicuously — to the cultures of Europe. It was argued that these cultures are gripped in a serious paradox that is perhaps without any exact historical precedent. Namely, that these are sex-repressive and sex-oppressive cultures, even while they sometimes do not appear to be so (cf., Foucault, 1988–1990).

On the one side of this paradox, these cultures appear quite "sexy." Sexuality — of a certain sort — seems to be on the surface of everyday life. Its images are readily accessible through the internet and other media (from seemingly innocuous advertisements for tight jeans, through buffed bodies in glossy magazines, to exhibitions of gang-bangs that can appear with the click of your mouse's button). Talk *about* sex acts is comparatively common and the paraphernalia of sexual activity (from birth control pills to dildos) are reasonably available. It is not that there is anything "wrong" with any of this. However, it must be recognized that what is purveyed in this fashion is a sort of commodified and commercialized "sex." It is a prime example of what some social theorists call "reification" — the mechanism by which a human process, perhaps even a sacred process, is treated as if it were merely a matter of things that are to be manipulated. The sexiness of contemporary culture does *not* mean that people are more readily able to listen to the voice of their embodiment. On the contrary, reification contributes to the mechanisms by which we become alienated from our embodied experience and this sort of "sexuality" becomes a compulsively obsessive mode of activity.

On the other side of the paradox, there may thus be a sense in which, as much or more than ever, the inherent sexuality of our embodiment is being systematically suppressed, repressed, and oppressed. Not only is there a backlash against the — somewhat mythical — "sexual revolution" of the 1960s, there are also other factors which contribute to the anti-sexual dimension of our contemporary acculturation. The backlash involves powerful forces that would compel our children into sexual illiteracy, leaving them anxious about any course of action other than abstinence, and keeping them ignorant of the methods of responsible sexual pleasure. These forces also promote the delusion that sexuality is dangerously addictive and they seek to "clean-up" both the public domain and our private lives. In this context, a compulsively phobic orientation toward sexuality is promoted. But in addition to this backlash, the cultural changes that have occurred in recent decades — including the apparent liberalization of matters around sexual behavior — may actually have exacerbated the mechanisms of reification that I just described. In short, there is little, if any, reason to believe that people are now less frightened of listening to the voice of their embodiment than they ever were.

This is *the paradox of sexification*, in which our contemporary culture has become *simultaneously both compulsively sex-obsessive and compulsively sex-phobic*. These compulsive tendencies react to each other and against each other — each tendency building on and against the other. Yet, at their psychological source, both arise from an ignorance and fearfulness toward the wonders of our sensual and sexual potential as human beings. Both are compulsive reactions against the power of our bodymind's inherently erotic potential, and both contribute to the mechanisms by which we become alienated from our embodied experience.

We can now ask where the community of bodymind therapists and the community of sexologists each stand with respect to this paradox.

As I comprehend the current situation, there is a serious problem facing bodymind therapy and somatic psychology. Ever since the paradigm of orgasmic capacity as the hallmark of health and wellbeing was advanced in the early twentieth century by seasoned clinicians such as Gross, Reich, and Balint, many bodymind therapists have been reluctant to acknowledge that their practices have anything very much to do with sexuality. To give a couple of examples of this, we might note that, in 256 pages, Hartley's otherwise excellent 2004 textbook titled *Somatic Psychology* has no more than seven sentences containing the words sex, sexual or sexuality (four of these references are in her reviews of historical material and one is in reference to abuse). Levine and Frederick's highly influential 1997 volume, *Waking the Tiger*, mentions sexuality exclusively in terms of the contribution of sex acts — meaning rape — to trauma. He does, however, mention in passing that "exaggerated or diminished sexual activity" can also be a consequence of traumatic developments (pp. 148–149). And on the psychoanalytic side, the terms sex and sexuality never even appear in the index of Anderson's 2008 anthology, *Bodies in Treatment* (which ironically is subtitled, *The Unspoken Dimension*). It seems that increasingly bodymind therapists wish to avoid speaking of sexuality, let alone exploring its significance in any detail — thus replicating the history of psychoanalysis by progressively avoiding the topic.

Obviously, there are notable exceptions to these generalizations. These exceptions mostly come from the Reichian tradition. However, the conclusion that the practitioners of bodymind therapy are often as fearful of sexuality as is the mainstream culture seems unavoidable. If one tracks the literature under this rubric, one notes a tendency for bodymind theorists to write about sexuality less and less as the twentieth century progresses, but also to write about the topic in less and less specific detail in relation to sexuality as an embodied experience. This, I believe, reflects the vulnerability of this professional community to the processes of sexification that have occurred within the mainstream culture. All too often bodymind therapists are succumbing to the sex-phobic tendencies that are ideologically evident in our culture's recent history.

The discipline of sexology is itself vulnerable to the culture's sexification. Sexology as *ars erotica* is an ancient discipline, as is known through such books as Ovid's *Ars Amatoria*, Vatsyayana's *Kāma Sutra*, and Kalyana Malla's *Ananga Ranga* (Kuefler, 2007). However, as a systematic science, sexology was launched in Europe at the very end of the nineteenth century and the beginning of the twentieth. There were many early contributors including Richard von Krafft-Ebing, Freud, Gross, and Reich. From its earliest years, it was concerned with understanding sexual minorities and with exposing societal hypocrisy. It was also an activist discipline concerned with protecting the civil liberties of these minorities throughout Europe, including, on Reich's part, an interest in protecting the sexual rights of adolescents (Reich, 1971a, 1971b). On the American side, although sexology as a discipline was not much in evidence, activism around sexual issues was conspicuous in this early period. For example, Victoria Claflin Woodhull was a suffragist, who became a charismatic advocate of women's rights, free love, and labor reform (Woodhull, 2005).

By 1919, Magnus Hirschfeld had founded the Institut für Sexualwissenschaft in Berlin. In those early days there was close collaboration between sexologists and the psychoanalytic community, and there was an interest in the inner processes of sexual experience. However, in 1933 the Nazis destroyed this Institute, burning its library. With the diaspora, the history of sexology underwent a hiatus. The discipline was to become reestablished on the North American side of the Atlantic with the work of Alfred Kinsey in the late 1940s and of William Masters and Virginia Johnson in the 1960s and 1970s (Bullough, 1994; Reiss, 2006).

In my view — and from the standpoint of somatic psychology — there are some significant problems with the sexological heritage of Kinsey and of Masters and Johnson, despite the importance of their different investigations. As valuable as they were, Kinsey's research interests were in large measure behavioral and demographic. He aimed to establish the frequency of various forms of sex act, as well as to be an active advocate for the rights of sexual minorities. Coming from a naturalscientific platform, the inner experience of sexuality for the human subject was scarcely within the scope of his interests. The path-breaking researches of Masters and Johnson conformed to the allopathic medical model. They were concerned to document the typical anatomical and physiological mechanisms of desire, arousal and climax. They had little regard for the meaningfulness of sex acts for the participant, nor were they interested in sexual attitudes beyond those of their own culture. Additionally, since homosexuality was considered pathological at that time by the American Psychiatric Association (and was until 1973–1974), Masters and Johnson were active in experimenting with conversion techniques designed to compel homosexuals into heterosexual behaviors (an enterprise that Kinsey would have detested). In short, the Masters and Johnson approach to sexual science could be said to be mechanistic and governed by assumptions about what "normal" sex might be.

Contemporary sexology — in the wake of Kinsey, Masters and Johnson — has generally continued to focus on sex acts, their frequency, their situational context, and their mechanics. In short, it has typically been behavioral in its orientation. This is certainly true of John Money's prodigious work, which has exerted a major influence over the field. Of course, there are some notable exceptions to this generalization, such as the remarkable investigations undertaken by Alphonso Lingis (1983, 1985, 1994, 1996, 2005). But the point here is that sexology has developed within an objectivistic or natural-scientific framework. It lost its original interest in the inner processes of sexual experience, and the fruitful alliance of the discipline with psychoanalytic explorations virtually ended with the diaspora. In this post-Kinsey period, most sexology has been married to the framework of allopathic medicine or that of cognitive behaviorism. The consequence of this is that sexology has failed to attend to the somatic psychology of human sexuality, let alone to develop what I am calling *somatic psychodynamics*. There has been far too little attention to the subject's experience of the sexuality of his or her embodiment, and far too little research on the subject's embodied experience of his or her sexuality. Instead of listening to the voice of our sexual embodiment, sexology has restricted itself to the investigation of behaviors and attitudes. One implication of this is that sexology has become somewhat detached from the remarkable research on bodily experience that is currently being undertaken within the disciplines of anthropology, sociology, and cultural studies. More generally, it must be concluded that all too often sexologists are succumbing to the sex-obsessive tendency that is ideologically evident within recent cultural developments — sexual mechanisms and sexual behaviors are extensively researched, but the inner experience of the sexuality of our embodiment all too often goes missing. In this way, the advocates of sexological science sometimes wind up contributing to the prevalent processes of our alienation from embodied experience.

A discipline of sexology without the holistic wisdom of somatic psychology is going to reinforce — in all those affected by its practices — the alienation from embodied experience that characterizes this culture. Sexology ends up being vulnerable to, and indeed contributing to, the processes of sexification that I described earlier. Conversely, a discipline of somatic psychology that avoids the focus of sexology is going to reinforce — in all those affected by its practices — the same alienation; I will return to this point shortly. What is currently needed is a new sub-discipline, which I will name *somatic sexology*.

The need for somatic sexology can be argued as follows. For a start, it is surely very odd that, in all that has been written about human sexuality, there has been almost no inquiry about the corporeal experiences that contribute to our awareness of the sexuality and sensuality of our embodiment. The literature on penile-vaginal intercourse is a prime example. There is much research on the anatomical and physiological mechanics of the act, much research on the different positions in which it may be undertaken or the effects of these positions, and much research on who has intercourse with whom and how often. But there is almost no research on the inner experience of the participants. The experiences of desire and arousal within the bodymind are exquisitely intricate and varied. Yet almost no attention has been paid to the topic. Similarly, it is known that climax can either be a localized spasm that affects only the pelvic area, or a full-bodied momentum of energy that initiates a temporarily altered state of consciousness. Yet, in the western tradition, research on this difference (between a climax that is detached from the entire bodymind, and one that rocks the entirety of our being) has scarcely advanced since the pioneering work of Reich (1980b) and Balint (1995) on the topic of genitality. Indeed, the most recent fourth edition of the *Diagnostic and Statistical Manual of Mental Disorders*, which is the handbook for all mainstream mental health practitioners, fails to distinguish ejaculation from orgasm in its presentation of male sexual difficulties — a failure that many Tantric practitioners view as preposterous. Penile-vaginal intercourse is a crucial human experience, celebrated by poets through the ages. It is crucial for reproduction, for heterosexual recreation, and for certain mystical practices intended for the transcendence of our egotism. It is regularly indulged by a very high percentage of the planet's population. Yet as an embodied experience, very little is known about it, and what is known has been contributed more by the products of the lyrical imagination than by scientific inquiry — orgasm as a spiritual experience will be discussed briefly in Chapter 16.

If further examples are needed, we might consider three. One would be the sensuality of the perineum and anal sphincter. Here is a fascinating region of the body that is highly erotic, yet scarcely mentioned in any scientific literature. There are some guidebooks for those who want to engage in digital-anal or penile-anal intercourse, and there are a few notable literary descriptions of the pleasures to be experienced in this region. But, to my knowledge, that is more or less all there is. Another interesting phenomenon is that of blowing or picking one's nose. Here are universal behaviors about which there is a small medical literature, which is entirely concerned with the physical impact of these activities. There is also a very small sociological literature concerned with the social codes that regulate nasal behaviors (Elias, 2000; Johnston, 2001). Yet given the universality of these behaviors, as well as the extraordinary sensitivity of the area around the columella and nares (not to mention the subtle connections that Tantric practices have discovered between the philtrum and the genitals or pelvic floor), one might imagine that there would be further inquiry into the experience of nasal activity. Finally, the sensuality of the feet and hands must be mentioned. Here too the erotic nature of these extremities has been seriously neglected in the scientific literature. While there is an extraordinarily detailed and informative body of literature — from East Asian, South Asian and Southeast Asian healing arts — on the subtle pathways by which the feet and hands are connected with every aspect of our embodiment, there is far less attention in the western literature to these matters, and almost no attention to the sensuality of the experiences derived from these extremities.

Somatic sexology — the marriage of somatic psychology and sexology — is greatly needed. This would be a human-scientific discipline, as contrasted with an allopathic, natural-scientific, or literary-aesthetic one. It would research, in a phenomenological or hermeneutic mode, the experience of our embodiment as a sexual or sensual experience — whether in pleasure or in pain — and it would bring into focus the way in which any profound therapy for the ailments of the bodymind also has to be a sexual healing. If the proponents of bodymind therapy continue to avoid the focus of sexology, not only will the discipline curtail its potential, but it will also contribute to a perpetuation of our alienation from the experiential body that characterizes our culture (and that the initiation of somatic psychology sought, at least in principle, to remedy).

Consider here messages that are conveyed to a patient by a bodyworker who addresses every aspect of the patient's embodiment *except* the genitals (cf., Painter, 1984). Not only are the penis, testicles, and vulva-vagina being treated as "special" and forbidden. In the very act of prohibition, all the forces of shame and guilt endemic to our upbringing are being reinforced. Moreover, if we take seriously the holistic nature of our bodymind and all the experiences that arise from it, how can healing occur if implemented under these conditions of partiality? The very act of exclusion of one region of the body from consideration must inevitably limit the impact that the healing process can have on the entire bodymind. It would seem obvious that any act of exclusion not only curtails the impact of the therapy, but also has the malignant effect of perpetuating the patient's alienation from the fullness of his or her embodiment.

This effect is the inverse of that achieved by some new age purveyors of what they call "Tantra." Such practitioners mistakenly believe that Tantric methods are "all about sex" and little else — failing to recognize that Tantra is first and foremost a holistic spiritual practice (Barratt, 2006). Emphasizing exclusively the energies generated in the first and second chakras (from the anus and perineum, through the genitals, to the upper pelvis), these misguided individuals fail to realize the "intent" of the first and second chakras. This "intent" is to be integrated in their connection with the third through the seventh chakras. Likewise, the "intent" of the crown or seventh chakra, as the peak of our energetic functioning, is to be integrated with all the energies of our genital region. In more common language, what this means is that cultivating the energies of the pelvic region merely contributes to an alienating procedure, unless these energies are drawn into, and aligned with, the heart and the head (Barratt, 2004a, 2006).

If somatic psychology and bodymind therapy are to develop as the holistic principles and practices that they proclaim, then not only does the pelvic region, including the anus and genitals, need to be addressed as much as the rest of our embodiment, but the discipline's congruence with somatic sexology needs to be acknowledged. This calls for a clear understanding of what human sexuality actually is.

A fundamental aspect of the sexification of our contemporary culture is the way in which "sex" is defined and discussed. In most instances, "sex" comes to mean intercourse, which almost always refers to penile-vaginal intromission. Even the etymological history of the word is noteworthy here. "Intercourse" comes into the English language in the fifteenth century from the Latin *intercursus* via the medieval French *entrecours*. It means "communication to and fro" or a "running between" of energy between two entities. Only at the very end of the eighteenth century does its meaning become narrowed to that of a sex act. This history is illuminating because the reification of sexuality in our contemporary culture is such that we generally do not think of penile-vaginal intercourse, or any other sexual intercourse, as a communication and as a running between, or exchange, of energies. Rather, we tend to think of it as "fucking," as one person "doing" another person, and as a penetrative activity. Interestingly, two possible derivations of the word "fuck" are the medieval English *firk* meaning to beat, and a proto-Indo-European root *bhau* meaning to knock or strike. It is surely telling that the connotations of intercourse are no longer that of a sacred exchange of energies, but rather of an act that has possible aggressive significance.

In earlier chapters, I have described how psychoanalysis began its history with an understanding of sexuality as the libidinality of our entire embodiment. As the twentieth century progressed, the notion of libido was retained as a term but emptied of its original energetic meaning. Instead, in the psychoanalytic community, sexuality came to mean "sex acts" of various kinds. Similarly, the western history of somatic psychology and bodymind therapies starts with the notion of sexuality as an ubiquitous energy — like Reich's theory of orgone — and devolves to the present situation in which many bodymind therapists want to assert that their work has nothing to do with sexuality, and indeed that bodymind therapy and sex therapy are two entirely different enterprises.

If sexuality is a circulation of energies within and around the entire bodymind, and a running of energies between two or more entities in any sexual experience, then somatic psychology and bodymind therapy need to embrace it as such, and to acknowledge its healing properties. And if indeed sexuality is optimally understood as a communication — as a movement of sensual information — then this understanding generates clinical and sociopolitical practices by which the mechanisms of reification and alienation will be challenged. Dissenting from the reifications of contemporary culture, we need to develop this understanding of human sexuality as something far more than — and occasionally different from — what are customarily designated as "sex acts." We need to retrieve and assert a viewpoint that is something like that of Efron's when he promoted the term "sexual body," adding to it the qualification that *the entire bodymind is sexual* (Barratt, 2005; Efron, 1985). Such a notion empowers us to surpass the dichotomy of compulsively sex-obsessive or compulsively sex-phobic discourse, which characterizes today's western societies. This brings us to the theory that I have developed of *the inherent polysexuality of being human*. This has three axioms (Barratt, 2005):

- The human bodymind is born with a polysexual potential for erotic pleasure; that is, we are not born to suffer for our erotic nature, nor is it necessary that this nature is constrained or curtailed, conditioned and restricted, by prohibitions and inhibitions.
- Typically, this potential is narrowed in the course of development as traumatic experiences constrain or curtail our bodymind's capacity

to experience erotic pleasure, and we become more alienated from our embodied experience.

- The incest taboo is the unique source of our fear of embodied pleasure, and in the course of development it typically is extrapolated to a multitude of anti-sexual and anti-sensual prohibitions and inhibitions.

We are not born into the world alienated from the pleasures of our embodiment, nor are we born with blockages in our ability to listen to the voice of our embodiment. Rather, we are born a holistic and integrated organism — without distinction of body and mind. Who can deny that the infant — the "bundle of joy" — is a package of free-flowing energy with neurons designed for an abundance of joyous pleasuring? Even an injured baby still lives fully in his or her body, without blockages in the circulation of energies. However, the powerful and expansive potential for sensual-sexual joy, which is our birthright, invariably becomes stifled and truncated through the procedures of our socialization and acculturation, which unavoidably involve traumatization (using this term in the broad sense to imply any experience which overwhelms us and thus effects a dissociation or alienation from our embodiment).

The polysexual potential of our erotic nature implies that we have the ability to manifest in our adulthood any and every pattern of sexual inclination ("lovemaps" as sexologists sometimes call them). We are not born with a predetermined program that stipulates that we will find our adult pleasures in this way but not in that way (for example, in the genitals but not in the anus, with men but not with women, and so on). There is no such thing as an infant who allows him or herself to enjoy the sensations of the genitals but not of the anus. There is no such thing as a heterosexual infant or a homosexual infant. There is merely an infant with an abundance of erotic potential, whose subsequent traumatizations will narrow the range of activities in which his or her pleasures are permitted (but this theory does *not* imply that this narrowed range is going to be experienced as a matter of choice or preference).

This theory of polysexuality accords with Freud's argument that we are each born with a wellspring of libidinality — subtle energies and sensations that move and circulate freely throughout our embodiment. This circulation is channelized or conditioned in the course of development, such that blockages occur. Our libidinal energies are no longer free-form and free-flowing. To greater or lesser extent, we all become alienated from our embodied experience. To reiterate this point: "at birth, our erotic nature is such that we have the potential to grow into adults who might experience every type of sensual and sexual pleasure that has ever been experienced by an adult human being ... yet the experience of our socialization and acculturation does not cultivate this erotic potential ... our 'maturation' involves the forceful diminution of this potential, and makes it the cause of our anxieties and our self-alienation" (Barratt, 2005, p. 92). This is — I believe — the basic wisdom of bodymind therapy. It implies that "sexuality" is to be understood as the holistic momentum of the bodymind's energies, and it implies that healing the bodymind must address these energies holistically.

It is to be anticipated that these arguments will evoke a certain amount of consternation in some practitioners, since their implications are trenchantly contrary to the prevailing *mores* or ideologies of our culture. After all, I am arguing that all healing of the bodymind is a sexual process, and that to deny this or to omit the genitals from the practices of healing is to perpetuate the very mechanisms of reification and alienation that bodymind therapy is designed to remedy. In this context, it seems important to discuss further what is meant by "healing" and by "health."

Central to the wisdom of somatic psychology is the wisdom that our bodymind heals itself, if given the opportunity to do so — an opportunity which is facilitated by addressing the blocks and obstructions it holds against its own healing processes. In Chapter 4, we considered healing processes in terms of: holistic discourse, energy mobilization, and appreciative connectivity. Pointing to the inherent healing properties of the lifeforce, I argued that healing is to be understood not as an avoidance of pain, nor as an avoidance of death, and not as a compulsory procedure of political or sociocultural adaptation. Ailments are blockages to the healing power of the lifeforce — its free circulation — and healing is the mobilization of the lifeforce as well as a presencing of our awareness of this power within and around us.

The definition of health — particularly in the context of sexuality — has become horrendously tied to the moralizing criteria of political and sociocultural ideology. For this reason, in *Sexual Health and Erotic Freedom*, I argue for a minimalist definition of sexual health as any mode of sensual expression that is undertaken in a way that is:

- Safe — which means that no physical or emotional harm is involved, and whatever risks there may be are deliberately minimized.
- Sane — which means that, whatever the activities involved, they are undertaken with awareness, and without coercion or compulsivity.

- Consensual — which means not only that all parties involved are fully free to give their consent, but also that all parties involved are of equal standing in their ability to consent.

Since incest does not seem to be possible without considerable emotional harm, and in any event rarely involves parties who are of equal standing in their ability to undertake the act with awareness, we can be quite confident in determining that it is an insane act that invariably damages the participants. However, perhaps more saliently, these three criteria caution us against the idea that any sexual act between participants from different generations and age groups, or between participants of unequal power (such as boss and worker, teacher and student, therapist and patient) can ever be healthy. And if not healthy, such interaction will never be a healing process.

These considerations are important because I am suggesting that bodymind therapy cannot conform to the prevailing *mores* or ideologies of a culture that systematically inscribes within us our alienation from the experience of our embodiment and from its wisdom. I will give three examples. Our social context dictates that therapists and patients should not touch, yet our wisdom demonstrates that touch is often intrinsically healing, and that touch of the genitals — the part of our embodiment that is most often associated with shame and guilt — might have an affirming and reconnecting function that is profoundly healing for some patients (Russell, 2003; Stubbs, 1999). Our social context dictates that therapists and patients should not be naked together, yet our wisdom demonstrates that body acceptance and appreciation is profoundly important for healing, and that working and playing in the nude — refusing to perpetuate the message that clothing is necessary because some aspects of the body are prone to shame and guilt — might also have an affirming and reconnecting function that is profoundly healing for some patients (Britton, 2005; Goodson, 1991). Finally, our social context dictates that therapists and patients should never be sexual in each other's presence. If by "sexual" arousal and orgasm are implied, then this is probably wise 99% of the time. Yet our wisdom also suggests that there might be circumstances in which sexual enactment contributes to a healing process. Anyone with knowledge of the healing practices that are sometimes accomplished by sexual surrogates and sacred prostitutes cannot deny this possibility (Brown, 2007; Keesling, 2006; Qualls-Corbett, 1988; Stubbs, 1994).

Let us follow up on the arguments presented in Chapter 13 and be very clear about one thing. Therapists who exploit patients for their own gratifications commit a travesty against the processes of healing. Patients with whom clinicians have enacted their own sexual desires, verbally or physically, probably never recover from the trauma and terrible abuse to which they have been subjugated coercively. Such enactments are unconsciously incestuous and, in a profound sense, always coercive, even if the patient appears to be a willing and desirous participant.

However, given the above considerations, it is clear that the future of somatic psychology and bodymind therapies will not be served by acquiescence to the injunctions of an insane culture — a culture that forces us to grow up absurd, to use Paul Goodman's words (Goodman, 1960; Gruen, 2007). Rather, the discipline needs to take a stand against a culture whose ideologies perpetuate the reified treatment of our embodiment and our alienation from its wisdom. But to take this stand, somatic psychology and bodymind therapy will need to acknowledge fully the inherent sexuality of being human, and to recognize the implications of this acknowledgement.

15
Oppression and the Momentum of Liberation

That there are political implications to the rise of somatic psychology and bodymind therapy seems inescapable — however much some of the proponents of this discipline might wish to avoid these implications for the sake of greater acceptance within the mainstream of psychology and the mental health industry. The discipline operates in a sociocultural and political context that cannot be ignored. In their path-breaking book, *Toward Psychologies of Liberation*, Watkins and Shulman (2008, p. 1) succinctly summarize our global situation:

> At the beginning of this new millennium, after hundreds of years of colonialism and neocolonialism, we cannot escape the disturbing fact that we live in a world where more than a billion people lack sufficient shelter, food, and clean water; where lakes, rivers, and top soils are dying; and where cultures clash and war, genocide, and acts of terrorism seem ordinary. Transnational corporations with vast reach and power control land, media, economies, and elections. Their policies are decided away from public view, in national and international arenas where the super-rich and super-armed preside. Economic globalization undermines much that is local and personal, affecting possibilities for housing, jobs, cultural expression, and self-governance. Such globalization has created a tidal wave of displacement, undermining families, neighborhoods, and cultures ... The psychological effects of deepening divides between the rich and the poor, unprecedented migrations, and worsening environmental degradation mark this era as one requiring extraordinary critical and reconstructive approaches.

The double question I wish to engage briefly in this chapter is as follows. What special insights from somatic psychology could help us understand how and why it is that human beings inflict such suffering on each other and on the planet we all inhabit? What could be the contribution of somatic psychology and bodymind therapy to the remedy of these issues? That is, what role could somatic psychology have in developing those "extraordinary critical and reconstructive approaches" that are so urgently needed for the future welfare of the planet and all its inhabitants? Watkins and Shulman show us how "all over the earth, innovative liberation psychologies are asking what kinds of psychological approaches might enhance capacities for critical thinking, collective memory, peacemaking, and the creative transformation of individuals, groups and neighborhoods" (p. 2). Indeed, nothing less than global cultural change is needed. So it is warranted to ask in what way the discipline of somatic psychology could have a vital role in the elaboration of these psychologies of liberation and thus contribute to radical cultural, political and economic transformation.

I believe this question is so momentous and of such crucial significance that it is far beyond the scope of this book to do more than adumbrate some directions for future consideration. In this chapter, I will try to touch on five interrelated arguments: first, that somatic psychology and bodymind therapy have unavoidable political implications; second, that liberation psychology needs to embrace the insights and methods of somatic psychology; third, that the structures of oppression are indeed encoded in each individual's embodied experience; fourth, that the commonplace definitions of health, especially mental health, need to be re-examined for their ideological underpinnings, and the notion needs be re-thought; and fifth that somatic psychology occupies a very special role in an epoch of epistemic shifting and global transformation.

It is important to sketch out the issues because, all too often, clinicians of every persuasion fail to appreciate the sociocultural and political context of their profession, as well as the unique contribution of their practices to the betterment — or to the worsening — of the general human predicament. Twentieth century psychology has operated on a Eurocentric model that (even in *social* psychology) tends to take the supposedly autonomous individual as its unit of analysis and to assume the immutably static condition of the social and cultural structures within which the individual functions. Typically, clinicians limit the scope of their investigations to the operation of this individual. However, in a critical sense, the autonomous individual is a fiction, despite this notion's centrality in western ideologies since the medieval era (cf., Wilson, 2004).

This model of psychological analysis is, in so many ways, a pervasive and profound mistake, which psychologies of liberation seek to counteract (Martín-Baró, 1994). If each individual's personhood is a distillation of history and culture, as has been demonstrated so extensively, including by the critical theory of the Frankfurt School (e.g., Adorno, 2001; Horkheimer, 1990; Horkheimer & Adorno, 2002), then to consider psychopathology (or "normality") solely in terms of personal reality is "a delusional repression of what is actually, realistically, being experienced" (Hillman, 1992, p. 93).

Somatic psychology and bodymind therapy have yet to meet fully the challenge of this insight. In Chapter 4, I discussed carpal tunnel syndrome as an example of the way in which the socioeconomic division of labor may leave a gross mark on the body's functioning. The social structures that compel some people to undertake repetitive movements for six, eight, or twelve hours a day — movements that are bound, sooner or later, to damage connective tissues — are not necessary for a society in which everyone has adequate shelter and nutrition. Rather, they are required for a society that is committed to capital accumulation and the aggregation of surplus wealth for the ruling classes. The production of this surplus requires the "efficiency" of a division of labor — a socioeconomic arrangement in which some will work as typists and some will plow the fields, and everyone will be alienated from the products of their labor. It also requires the ideological reification and commodification of our embodiment. The message of strained and damaged connective tissue is, in an important sense, a voice of dissent against the injustices of compulsory wage slavery in a society, and a global economy, that is organized for the material enrichment of those who are already rich. Yet how many of those suffering carpal tunnel syndrome experience their pain in this manner — or instead, do they chide themselves for "weakness," feel resentment toward their wrist and hands, lose sleep and accumulate anxieties, experience an increasing sense of despair? The point is not that every act of healing has to be a lesson in sociopolitical consciousness; but that a fully scientific understanding of any ailment of the bodymind needs to include the interpretation of the ailment in its social, cultural, political and economic context.

If this is the challenge that somatic psychology needs to meet, it must be said that it is currently met only ambivalently. As I have already pointed out, psychoanalysis began as a radical and subversive discipline and can be understood as the harbinger of postmodern impulses (Barratt, 1993); many of its pioneers were activists of a leftist or liberal democratic persuasion, committed to issues of social justice (Danto, 2005). However, after the diaspora, a radical vision of social change ceased to characterize the psychoanalytic community; too many psychoanalysts abandoned the subversive implications of their discipline in exchange for financial privilege, social prestige and conformity (Altman, 1995; Cushman, 1996; Jacoby, 1975, 1983; Oliver & Edwin, 2002). A similar retreat might come to characterize the fledgling community of bodymind therapists, unless it holds to the radical implications of its own discipline. The explicit political vision of those who contributed to bodymind therapy in the early years of the twentieth century, such as Reich, already seems less in evidence in the latter years of that century — although there are notable exceptions. This is perhaps because too many somatic psychologists and bodymind therapists have become over-anxious in their pursuit of acceptance within mainstream psychology and the mental health industry. This goal, as I hope to show, is thoroughly misguided. Somatic psychologists and bodymind therapists may conceptualize their discipline in terms of the individual's functioning, but this conceptualization is something of a deception that will neither serve the interests of patients nor the future of the discipline.

That this sort of ideologically circumscribed investigation of individual functioning is deceptive is precisely why Ignacio Martín-Baró, the initiator of liberation psychology, called for a re-visioning of the goals, the epistemology, and the praxis of psychology (Martín-Baró, 1994; McLaren & Lankshear, 1994; Watkins & Shulman, 2008). According to Watkins and Shulman, this call is echoed in diverse ways by the depth psychologies of such writers as Albert Memmi, Frantz Fanon, Aimé Césaire, Paolo Freire, Gloria Anzaldúa, Aurora Morales, Susan Griffin, Chela Sandoval, Elizabeth Lira, Enrique Dussel, William Du Bois, James Cone, and Audre Lorde. This list is far from exhaustive. Additionally mention might be made of the important philosophical work of Enrique Dussel (1985), which has clear implications for our vision of embodiment. It cannot be said that western psychologists have, as yet, greatly attended to this call, although there are some promising exceptions (e.g., Alschuler, 2006; Fox, Prilleltensky & Austin, 2009; Prilleltensky & Nelson, 2002; Parker & Spears, 1996; Sloan, 1996, 2000; Teo, 2005; Tolman, 1994).

Liberation psychology addresses the connection between awareness (consciousness and the dynamics of the unconscious) and sociocultural or political and economic forces. However, liberation psychology has not yet come to terms with the bodymind perspectives of somatic psychology (Fanon's discussion of the "epidermalization" of oppression is something of an exception to this generalization). As such, liberation psychology is sometimes in danger of limiting its ability to explain people's resistance to social and cultural changes from which they themselves would benefit.

Traditionally, those who have critically examined the societal mechanisms of ideological reproduction have limited the scope of their investigations to cognitive — attitudinal and motivational — factors. Although the term "ideology" was coined in the late eighteenth century by Antoine Destutt de Tracy, whose philosophy was sensualistic (he developed a distinction between active and passive touch), subsequent usage focused on the transmission of representational mental activity. This can be said of the tradition of ideology critique that runs from Karl Marx to such twentieth century commentators as Antonio Gramsci, Karl Mannheim, György Lukács, and Louis Althusser. Although varying in emphasis and orientation, in the writings of these critical thinkers, ideology reproduces the social order in a manner that is mostly cognitive. Ideology is defined both as a set of ideas with specific content and also as a form or structure by which, and within which, ideas are generated. Ideologies govern both the role of ideas in the individual's social interactions and the role of ideas in the structuring of organizations and institutions. This focus on ideology as a feature of mental representations is also true of psychoanalytic efforts to its investigation (Barratt, 1985).

Yet the ideological reproduction of oppressive social structures is mediated by the entirety of the bodymind, and not just by the cognitive functions of ideation, conation, and reasoning (along with the affect that accompanies them). The individual's assimilation of ideological functions is, in many respects, the prototype of self-alienation. It is the production — in which actions become internalized as beliefs and emotions — of "subjectivity without a subject" (Macherey, 1998, 2006), which can be considered a process of "becoming-other" (Deleuze, 1995, 2005; Deleuze & Guattari, 2004, 2009). None of the authors I have mentioned would argue that alienation is exclusively a mental or representational event — even if their focus has been on representational activity. Rather, alienation impacts the entire functioning of the bodymind, immobilizing its energies and keeping the subject divorced from the awareness of its own desire (Barratt, 1993).

Somatic psychology potentially has much to contribute to understanding the processes of interiority or internalization by which the mechanisms of oppression become those of suppression, repression, and dissociation, within each individual. It also has much to contribute to understanding how physical damage to the human body results from oppressive social and cultural structures, as well as how the psyche responds to such damage. That is, somatic psychology is uniquely positioned to empower our understanding of the procedures by which an individual's maturation and adjustment as a "normal citizen" contribute to the ideological reproduction of the cultural and socioeconomic system within which he or she matures and becomes adjusted.

It is precisely somatic psychology that can instruct us as to why the individual's assimilation of ideological functions is so pernicious. Somatic psychology is the discipline that can understand why ideology is so intractable in relation to the reasonable arguments that can be made for much needed changes to our social and cultural structures — why the obstacles to urgently needed cultural transformation often seem so recalcitrant. It is precisely because ideology is encoded in the holistic functioning of the bodymind that, as humans, we are so often resistant to progressive social change (Barratt, 2009a). Let us briefly consider the various processes by which oppression becomes internalized within the bodymind of the individual.

This interiorization of oppression and our reactivity to oppressive structures are not only multifaceted but occur on many psychological levels. Although this matter is side-stepped by most of the theorists involved in ideology critique and liberation psychology, the individual's responsiveness to incest taboo is deeply relevant to the way in which individual's are primed for the assimilation of ideology in terms of sociocultural prohibitions and inhibitions. As I suggested in Chapter 13, the taboo is a "deeply-wired" necessity that safeguards the possibility of our sanity. As was stated previously, it engraves the boundary of reflective consciousness within the individual's psyche, for it establishes the "repression barrier," which demarcates the boundary between the thinkable and the reality that is unthinkable — the distinction between the acceptable realm of conscious-preconscious thoughts and feelings, and the forbidden realm of thoughts or feelings that are suppressed and then repressed into unconsciousness. As the psychological prototype of all subsequent prohibitions and inhibitions, what is forbidden is interpreted and extrapolated variously by different cultures and social groupings. It becomes the root force behind the perpetuation of ideologies.

That the incest taboo ultimately fuels the compulsive quality of all other prohibitions and inhibitions is peculiarly evidenced by the history of individuals from elite groups, who have deluded themselves into believing that they are "above" the taboo itself. The fact that leaders of dominant social groups — emperors, royalty, and presidents — can grandiosely fantasize that they themselves are "above the law," and the fact that such grandiose flights of imagination eventually undermine the reality of their dominion, actually points to the power of this law over both themselves and those who are their subjects. It also points to the function of this "law of laws" in perpetuating the systems of social domination. Even Freud is alleged to have flirted with this proposition that some people might not have to comply with the incest taboo in his unpublished correspondence with Princess Marie Bonaparte (who, significantly enough, was anorgasmic with vaginal intromission, and tragically underwent no less than two unsuccessful Halban-Narjani surgeries to relocate her clitoris closer to the vaginal introitus). Compliance with the incest taboo is necessary, but compliance with all the rules and regulations, that derive their psychological force from it, is not. Moreover, as I will shortly suggest, obedience to prohibitions and inhibitions always requires a loss of attention to the voice of our embodied experience — which is why the praxis of becoming aware of our embodied experience may have revolutionary potential.

It is the extrapolation of the incest taboo, as the "boundary of boundaries," into all manner of rules and regulations — prohibitions and inhibitions — that generates all the forces of shame, guilt, anxiety, fear, and malignant affiliation which make possible the transmission of ideologies. These rules and regulations are not held as abstractions on the level of mental representation; more powerfully they are encoded within our embodiment, and compliance with them requires that embodied experience be ignored.

In various ways, the incest taboo is always inscribed within the intricacies of the bodymind's functioning, including the circulation of its subtle energies, and involves the loss of our capacity for listening to the impulses of embodied experience. So too are its derivatives. Sometimes the taboo is even marked visibly on the body — as was discussed in Chapter 13 with reference to my discussion of the Ganesha myth (Barratt, 2009a). However, other boundaries, extrapolated from the incest taboo, also cause fragmentations in the holistic integration of our embodied experience. Some of these fragmentations or energetic blockages are subtle, some are gross. On a more subtle level, the conveyance of prohibitions and inhibitions manifests throughout our embodiment in blockages to the free-flow of energies, the constriction of our libidinality (cf., Marcuse, 1987). This implies that our potential for the awareness of embodied experience is constrained, conditioned or curtailed, which is central to the transmission of ideologies. It also implies a general loss of psychic energy.

The latter is an important issue for the mandate of liberation psychology, since it enables us to understand why those who suffer social injustice and cultural oppression so often seem to face their life's circumstances with fatalism or anomie — their capacity to imagine and create their circumstances differently becomes constricted. This has been somewhat explored in terms of the psychopolitics of abjection (cf., Kristeva, 1982). It is not only that those who are oppressed are frequently physically debilitated and politically disenfranchised; it is also that psychologically the interiorization of ideological tenets produces a ubiquitous sense of despair and disempowerment (cf., Marcuse 1971, 2006).

It can be seen here how somatic psychology might offer much to those interested in liberation of our human potential, and much of what is offered comes via its understanding of traumatization.

In general clinical practice and in the popular imagination, it has been a frequently committed error to believe that traumatization can be externally defined (which is not to deny that there are events that would traumatize any and every participant). Traumatization needs to be defined as any event or persisting circumstance that overwhelms the individual's capacity to process this experience cognitively, emotionally, and — especially — somatically. Traumatization is thus, by definition, an experience that threatens our bodymind's entire organization by causing a rupture in its capacity to assimilate and accommodate the embodied experience (such that it might take our organismic functioning to a new level). In this sense, traumatization freezes our psychic energy. Thus, under-stimulation, over-stimulation, and conflictual stimulation, can all have a traumatic impact on the individual, and what is "traumatic" will always be relative to the individual's pre-existing constitution, and will differ at each phase of our personal growth. What might be experienced as trivial to one individual may be profoundly traumatic to another.

Elsewhere I have argued that, when the stimulation constituted by an event or circumstance disrupts the capacity of the organization of the individual's bodymind to process the experience, the entire organization "faces" — in some sense — an intimation of its own annihilation, as if in a concrete experience of the "deathbound" condition of our subjectivity (Barratt, 1993, 1999). This is somewhat similar to Lacan's notion of the register of the "Real" (Lacan, 1972, 1977). In this sense, a "traumatic experience" always provides us with an intimation of the abyss that is within each of us — an encounter of the emptiness that is within, and an "unthinkable" experience of our own inherent "deathfulness." This brings about a closing-down or freezing and fixating of the flow of psychic energy — the lifeforce that runs within and through us from the cradle to the grave — that was so abundantly available to our pre-traumatized awareness.

In addition to the episodic experiences of traumatization that so many of us suffer — the physical and emotional assaults and abuse — it needs to be recognized that systems of social, cultural, political or economic oppression are traumatizing for every individual within them, in a multitude of ways that are not always so acutely visible. Whereas the sequelae of assault and abuse have been extensively investigated by somatic psychologists (e.g., Levine, 2008; Ogden, Minton & Pain, 2006; Siegel, 1999, 2007), the persistent and pernicious effects of participation in oppressive structures have been less discussed. The most outstanding example is the way in which social injustice and the ideologies of supremacy (the ideology and praxis of domination-subjugation discussed in Chapter 2) impact the bodymind of the victim, the perpetrator, and the bystander. This is *not* to imply that it affects them alike. Let us address each of them in turn, as has also been ably accomplished by Watkins and Shulman (2008).

For example, Memmi (1991, 2000) and Fanon (2005) provide a brilliant analysis of the psychology of those who have been victims of racism and colonialism (as one major locus of oppression). They document the smoldering anger that has to be suppressed and repressed, corroding the health of the bodymind, and frequently getting translated into lassitude and passive acquiescence with the injustice to which the victim is subjugated. That the ideologies of racism and colonialism become deeply encoded within the victim is further documented by the fact that removal of the structures of oppression does not immediately heal the interiorized traumatization (Fanon, 2008; Memmi, 2006). This is surely because the oppressive force of racist and colonialist structures becomes translated into energy blockages, and a closure of awareness, within the bodymind of those who are its victims.

It is not surprising that Fanon, in his earlier and more notorious writings, advocated revolutionary violence as an action that would heal the ailments of its participants, presumably by releasing the blockages within them. There are, of course, arguments against this mode of "healing," not least being the proposition that the enactive expression of hatred may be a release in one location, but only causes the accumulation of further hatred in some other location of the cosmos (Dalai

Lama, 1999; Gandhi, 2002). Later, Fanon wrote "today, I believe in the possibility of love" — his intent now being to identify its blockages, which he calls "perversions," and to explore methods of empowerment that would liberate those who suffer such ailments. This raises questions about the complexity of personal change that might incorporate both empowerment and forgiveness (cf., Derrida, 2001; Minow, 1998; Tutu, 1999). This is a matter to which somatic psychology makes a crucial contribution because forgiveness, if it is not to be an inauthentic compulsion based on the superego's forcefulness, has to involve the entirety of the bodymind. It is not that social and cultural ills can be cured by individual therapy; but it is the case that somatic psychology can help us explain why so many victims are deeply affiliated to the mechanisms of their victimhood, and this understanding shapes the possibilities of remedy.

On the side of the perpetrators, it is an error to imagine that the structures and activities of oppression do not also impact their embodiment deleteriously. Emotional numbing, loss of psychic energy, and anorgasmia (as exemplified on the male side by the ability to ejaculate but not to enjoy a full-bodied orgasming) are just a few of the common effects of living on the side of the oppressor. As Baldwin (1961) expressed it, "one cannot deny the humanity of another without diminishing one's own."

At one vivid extreme, we might examine here the psychology of killing. Grossman's military researches suggest that humans have a powerful inherent resistance to taking life, particularly the life of other humans. Specific experiences have to occur for this natural aversion to be broken; so, for example, military training has to condition recruits to overcome this resistance (Grossman, 1996). Related investigations show how, once the aversion is overcome, an "addiction" to killing can apparently develop, in which the subject becomes deeply attached to the "rush" of bloodshed and begins to find no other activity quite as arousing or gratifying. Clearly, what is at issue here is no mere shift in attitudes and values, but rather a traumatization of the entire bodymind and the consequences of this bodymind's efforts to heal itself by the repetition of the enactment.

Although Grossman's researches specifically address the complications of learning to kill, the dynamics apply equally to those who torture, assault and abuse, the "other." To implement these actions, the perpetrator has to block the natural capacity for empathy with the other. Processes of doubling (a form of splitting in which the participant can be a decent person in one context, and a barbaric evildoer in another), disavowal, derealization (in which an entire segment of experience is treated as "unreal") and generally diminished subjectivity, all accompany the bodymind transformations required to commit such actions as torture, assault and abuse. This is well documented in Lifton's studies of Nazi doctors, which unfortunately did not delve much into the somatic profile of these men (Lifton, 1986, 1993), as well as other researches into the psychic functioning of police torturers, death squad members, and rapists (e.g., Huggins, Haritos-Fatouros & Zimbardo, 2002; Zimbardo, 2008).

At a less vivid extreme, we might consider the everyday treatment of the "other" by those whom social and cultural structures assign to the dominant groupings. Colonialism, for example, decivilizes and brutalizes the colonizer (Césaire, 2001). The motif of domination-subjugation, which is ubiquitous throughout modern western culture, renders the "other" as an entity of no consequence other than to be dominated. For example, to implement the mechanisms of domination over third-world economies, the colonizer, the neo-colonialist, the racist, the international investment banker, or the CIA agent, all have to blinder themselves to the far-reaching implications of their activities, and to numb themselves to the atrocities with which they collude (e.g., Perkins, 2005). Although the available literature stresses defense mechanisms such as denial, projection, and rationalization, the effect of these blinkered and benumbed psychological operations is also somatic. These mechanisms by which domination over the "other" is maintained are evident in every arena. Women, children, and nature are traditionally treated as "other" — there to be mastered and controlled such the benefits they offer may be possessed (Griffin, 1996).

The motif of mastery and domination — control and exploitation — comes to characterize what might be called, following Susan Griffin, the erotics of everyday life (cf., Certeau, 2002; Griffin, 1996; Lefebvre, 2008). More accurately, participation in the activities of domination de-eroticizes everyday life. It closes down the bodymind's receptivity to what Martin Buber (1971, 2002) called the "interhuman," and Levinas (2005) called the "humanism of the other." It transmutes the connectivity of belonging into the maneuverings of possession, the relations of being into those of having and doing (cf., Fromm, 2005; Hyde, 1983; Kohák, 1984).

These same detrimental effects to the subjectivity of the perpetrator apply in large measure to the psychology of the bystander (as distinguished from the engaged witness), as well as all those whose daily life involves encounters with trauma (cf., Henry & Lifton, 2004). Watkins and Shulman adroitly discuss how "psychically being a bystander to injustice and violence breeds disconnection, passivity, fatalism, a sense of futility, and failures in empathic connection" (2008, p. 65). In this process, intelligence becomes severed from the life of the emotions and the imagination; the bodymind's energies become calcified. The processes of globalization position everyone who is on the side of the western world — especially those who are white, male, heterosexual, and affluent — as perpetrators or bystanders to greater or lesser degree. The price of privilege, which invariably implies collusion with the forces of oppression, is always a benumbing attenuation of access to the free-flowing source of the bodymind's erotic energies; it is a closing down of the heart in relation to both the sensuousness and the suffering of the world (Hillman, 1992). Without even any awareness of the extent to which awareness has been stifled, the bystander opts for that "seasonless world" where we may laugh but not all of our laughter, and weep but not all of our tears (Gibran, 1969, p. 12).

If these are some of the ways in which the structures of oppression are marked within each individual's embodied experience, then the notion of health, especially mental health, needs to be critically reconsidered. In Chapter 4, we discussed the authentic nature of healing, which is not an avoidance of pain, nor even necessarily its palliation, not an avoidance of death, and not a procedure of political or sociocultural adaptation. Genuine healing involves freedom and presence, both of which require our potential to be aware of our embodied experience. For example, in Chapter 14, I suggested that erotic activity is inherently healing and healthful, that sexual health means engaging our embodied pleasures fully and freely, but with awareness. I also argued that awareness, in the context of sexual health, implies that all such activities be safe, sane, and consensual. Similarly, the prevailing notion of general health as merely the absence of disease or injury needs to be critically revised. Although such a discussion is beyond the scope of this book, I want to emphasize how the field of mental health has been captivated by the ideologies of *adaptation* and *maturation*, as well as to indicate how these ideologies contribute to the perpetuation of oppression, rather than to our liberation from it.

The lamentably thoughtless notion that "health" entails the individual's ability to fit into the organization of the dominant culture and of the ruling social order serves admirably the ideological aims of the ruling-class in any oppressive social system. In this sociocultural organization, "well adjusted" and "mature" individuals are those who are markedly alienated from embodied experience and the wellsprings of their desires. This has been variously discussed as the "pathology of normality" (cf., Devereaux, 1980; Foucault, 1988; Goodman, 1960; Gruen, 2007; Laing, 1983; Mullen & Laing, 1996; Reich, 1974). Adjustment is viewed in positive terms as the ability to operate successfully within the dominant culture and the ruling social order; maturity often implies an acceptance of the "fact" that one's social and cultural circumstances are not open to change, or only minimally so.

Emotionally, "well adjusted" and "mature" individuals have to be somewhat closed down, their responsiveness and empathic potential diligently constrained, and their potential for imagination and creativity somewhat curtailed. Somatically, such individuals have to become somewhat insensitive or desensitized, their awareness of embodied experience closed down (Barratt, 2005). This has little to do with "health" as conceptualized by allopathic physicians, for this closed-down condition of reduced sensitivity to embodied experience is as much characteristic of top athletes as it is of those who are physically challenged (indeed, often the intensive training required of professional athletes requires that they cultivate a higher degree of insensitivity to the voice of their embodied experience). Anxiety, shame, guilt and fear contextualize the sexual experiences of those who are "normal," resulting in reduced genital sensitivity, inflexibility of the pelvis and spinal column, as well as thoracic constriction. These are the physical requisites for intense orgasmic experience, which explains why full-bodied orgasming is rarely — if ever — integral to the normal person's life experiences. To become "normal" one must become erotically numbed. In short, "normality" within the law and order of an oppressive sociocultural system requires that one is "out of touch" with the lifeforce flowing within — in large measure, the condition of adaptation and maturation is that of the walking wounded. This is the ideological tyranny of "appropriate behavior" (Barratt, 2005, pp. 49–56).

To transcend the conventional definition of "health" would be to return to Freud's twin criteria — *to be able to love and to be able to work.* To these, I would want to add — in the spirit of postmodern impulses — *to be able to play.* Yet each part of this trinity is problematic. In our contemporary world, what is called love is too often conflated with attachment; it is embedded in a network of obligations and expectations (I will return to this point in the next chapter). What is available for work is very often alienating — the division of labor requiring a seriously imbalanced exercise of our abilities. What is called play is too often confused with games — many of which are an anathema to the whimsical spirit of genuine playfulness. However, if these three criteria could be emancipated from their current context, they might provide a truly liberating image of health. Such a conceptual emancipation would require a considerable labor of ideology critique.

In an important sense, the practices of somatic psychology and body-mind therapy are intrinsically a labor of ideology critique and need to be comprehended as such. In the labor of ideology critique, it is immediately evident that there are three very forceful images of the body prevalent in our contemporary culture. Each of them represents the ideological reification and commodification of our experience of embodiment, and each needs to be critically deconstructed by somatic psychology. Each of them does untold harm to our experience of our selves. Here I will mention them only briefly. They are as follows:

- The Media Ideal. Specific cultural ideals of what a body is meant to look like, and how it is meant to be, are extensively and intensively conveyed by the media. The media promulgate images of attractiveness that are pronouncedly superficial, often unrealistic, and frequently unhealthy. Far from supporting us in the adventure of listening to the experience of our embodiment, the media encourage us to turn our attention outwards, to look at other bodies, and then to compare, contrast, and compete. Programming on television, on the internet and in the print media directly and indirectly promote body modification and augmentation, accumulating to each individual's lack of acceptance of their embodiment as it actually is. The glamorized body presented in these media reinforces bodily dissatisfaction and self-hatred in "ordinary" people. From plastic surgery that mutilates the body to sports that require the cultivation of entirely imbalanced and disproportionate bodies, the media exercise a quite malignant influence on our experience of embodiment.
- The Medical Ideal. It is difficult to criticize a profession that is designed to help us retain the functionality of our anatomical and physiological endowment. However, the objectivistic platform of allopathic medicine (in which each individual's body is seen as a cadaverous system of anatomical and physiological structures and mechanisms that happens to be alive) does not support the practice of listening to our embodiment as an experientially meaningful voice. The medical mindset necessary for physicians also pervades the public's attitude toward our embodiment in a way that is frequently far from benign.

The objectivistic treatment of the body as a machine that we possess — but that is ever likely to breakdown and become a liability — reinforces the processes in which we become alienated from the experience of our embodiment.

- The Economic Ideal. Bodies are also units that are required to fill specific positions in the economic relations of production. In Chapter 4, I alluded to the way in which our society places a premium on each individual becoming a "productive citizen," who can earn a wage — or support someone who earns a wage — and contribute to the profitability of the capitalist system (that is, the profitability of this system for the dominant socioeconomic class). In a sense, this generates the attitude that a body is only as good as its capacity to deliver surplus labor (the portion of one's labors that typically accrues to someone else's profit). The economic value of the body drives the conduct of healthcare within the medical model — the point of "health" is not so much the patient's happiness, but rather to get the patient back into the workforce, and/or to minimize the extent to which his or her ailment becomes an economic burden on the system. This was exemplified by my earlier discussion of carpal tunnel syndrome. The medical and economic attitudes toward the body are in turn reinforced by the ideals advanced in the media.

In short, these three ideals are complicit and their interacting influence on our contemporary experience of embodiment needs to be challenged and critically deconstructed.

If psychology is not to operate merely as a functionary of the dominant culture and prevailing social order, then it has to become a critical discipline occupied with these multiple ways in which ideologies condition and constrain our human potential. Somatic psychology could have a crucial role in this epoch of epistemic shifting and global transformation, for it is the discipline that can bring into focus how the structures of oppression are lodged within the bodymind of each individual. This is the common ground between the call for a re-visioning of psychology in the interests of liberation and the rise of somatic psychology or bodymind therapy. These disciplines contribute methods of praxis for modes of individual transformation that have wide-ranging repercussions for change on a social and cultural level. The human capacity to abuse and exploit others — our proclivity for discrimination, injustice, violence — surely requires our underlying disconnectedness from the awareness of our sensual and sexual bodies. The processes by which we can reconnect with the awareness of our embodied experience have far-reaching implications for cultural, social, political and economic change. Somatic psychology may well be destined to contribute profoundly to the psychology of liberation; this is one of its major challenges. And with this, it is also called to contribute to the realization of our spiritual potential.

16
Bodily Paths to Spiritual Awakening

According to the little known Gospel of Judas Thomas, Jesus of Nazareth once said, "Whoever has come to know the world has discovered the body, and whoever has discovered the body, of that person the world is not worthy" (Thomas, 1992, verse 80). Given the surrounding text, it seems likely that the "world" to which Jesus refers has two different senses in this passage. The world leading us to discover the body might well be that of the beauties of the natural environment and the marvels of our planet; the world not worthy of the person who has discovered the wonders of the body is that of human culture, with all its political, economic, and ideological machinations. Similar quotes can be found in the hidden teachings of great spiritual leaders from the Judaic and Islamic traditions, and even more abundantly in teachings from every tributary of the Dharmic and Taoic spiritual traditions.

Somatic psychology and bodymind therapy lead us back to the awareness of our embodied experience — proceeding against the images and concepts of the body propounded by cultural media, objectivistic medical sciences, and capitalist economies. In this context, we have to confront the challenge involved in acknowledging that the processes of returning to the awareness of our experiential embodiment are essentially a *spiritual* practice.

This chapter briefly sketches three ideas that I believe somatic psychology has to address: first, it will be argued that somatic psychology cannot avoid, and should not try to avoid, the charge of being a spiritual discipline; second, some notes as to how somatic psychology has fundamental relevance to the coordinates of secular spiritual practice will be presented; and third, it will be suggested that somatic psychology might benefit from the teachings of those mystical traditions that hold the universe to unfold within the body.

From a political standpoint, there is a sense in which listening to the voice of our embodied experience is a subversive act. It is a refusal to treat the body as a conceptual object or thing — the treatment accorded it by western culture throughout the modern era. In the act of responsive listening, the body is no longer merely an object of calculated attention, no longer material for instrumental manipulation. Rather, the body becomes a dialogical partner in the processes that constitute our being-in-the-world. Although there is a certain sort of estrangement between partners in dialogue, this ontological relation is profoundly different from the epistemological dichotomy of subject and object (as has been extensively discussed in the philosophies of hermeneutics since Heidegger). Estrangement is quite different from alienation in its recognition and appreciation of difference (Barratt, 1993). The act of listening to the voice of our embodied experience overcomes the alienation from our embodiment that is established in the course of our socialization and acculturation. It takes the reified or calcified state of alienation and mobilizes it into a lively dynamic of estrangement that might be called our *being-in-process* (cf., Barratt, 1993; Kristeva, 1975, 1984). In this sense, the agenda of somatic psychology and bodymind therapy transports us both beyond our captivation in the objectified and idealized body of the media, of medical sciences, and of capitalist economies, and beyond our entrapment in the dichotomy of *res cogitans* and *res extensa*. Listening to our experiential embodiment — and thus conveying our subjectivity beyond both the thoughts enunciated by our chattering mind and the physical mechanisms or "thingness" of our body — liberates us from our own alienation.

But our act of releasing ourselves from an alienated relation with our own embodiment is not only a procedure with social and cultural implications; it is also a restorative spiritual event. It is a revitalizing process of reconnection with the lifeforce within us, and in this sense it is holy (Otto, 2004). The contemporary challenge for somatic psychology is to be very judicious — which is not to say, circumspect — in the way in which it addresses the question of reconnection. With what is it that we are reconnecting? How this question is broached depends in large measure on the position one takes around the issue of subtle energy systems (which have been discussed several times in earlier chapters).

As has been indicated, it is possible to produce a sort of objectivistic "somatic psychology" that is *about* the body. This is quite different from a psychology *of* the body, which addresses and listens to the voice of our embodied experience. Similar to the mandate of psychosomatic medicine, an objectivistic psychology about bodily matters would address only issues involving the connective tissues — the observable anatomical structures and physiological functions — for the way in which they impact issues of cognition, emotion and motivation. In such a psychology, awareness of the body's voice would not be in question; rather, manipulative treatment of bodily attributes in order to modify their impact on mental life would be the intent of the discipline. There are several forms of bodywork that operate in this manner, and their mode of operation is concordant with that of psychosomatic medicine, rehabilitation psychology, sports psychology, and similar disciplines. However, if this is how somatic psychology is conceived, the scope and prospects of the discipline will be severely limited.

Most practices of somatic psychology and bodymind therapy operate on a notion of embodiment that is significantly expanded and divergent from the allopathic vision of the body as a complex set of interrelating systems of anatomy and physiology. In one way or another, most of these practices make reference to an esoteric dimension of our being-in-the-world. This is the body of subtle energies that are liminal in the sense that they are betwixt and between, "in but not of" the ordinary body of flesh and blood. This esoteric body has been described in multiple ways. It is the body of libidinality, orgone, prānā, chi, or *spirit*. It is sometimes referred to as the "breath body" — although it must be noted that "breathing" is here considered as far more general than the movement of air in and out of the lungs, such that it might become meaningful, for example, to talk of breathing through the perineum. This is the body of the lifeforce, the Bergsonian *élan vital* that is the brio or kinesis of life itself — its energies are life's longing for itself.

In Chapters 10 and 11, we reviewed some of the salient features of these subtle energies. Three features are particularly important. First, that blockages or obstructions to the movement of these subtle energies interfere with the bodymind's potential to heal itself; addressing such blocks and mobilizing these energies thus becomes the primary way of facilitating the healing process. Second, that these subtle energies circulate not only within, but also *around* the ordinary body of gross physical structures, not necessarily following the discernible histological pathways of neurons, glands or blood vessels; the movement of subtle energies thus connects, and renders interdependent, the individual's embodiment with the entire universe. Third, that the flow of subtle energies is intimately connected with the intentional realm of the imaginal. "Prāna goes where intentionality goes" is the Yogic-Tantric formulation of this phenomenon, which implies that our entire embodiment — and the universe around us — is, in an important sense, a creation of "mindstuff," and makes it not only meaningful but wise to consider the voice of our experiential embodiment. These features of the subtle energy systems that pervade our being-in-the-world make nonsense of the dichotomy between mind and body. They also suggest how subtle energies are, under another description, the foundational process and liveliness of our desire — our spirituality incarnate.

If we are open to the existence of esoteric energies, it seems unavoidable that we need to embrace fully the notion of somatic psychology and bodymind therapy as an *existential and psychospiritual discipline*. We need to relinquish the theological tradition that tells us we are bodies with souls, and embrace fully the notion that we are *spirited bodies* (Murphy, 2006). Both the theory and the practice of somatic psychology not only prompt a reconfiguration of the subject/object split and a transcendence of the mind/body dichotomy, but also inspire us to challenge the traditional division between the sacred and the secular. This discipline is not only a potential psychology of liberation in the social, cultural, political and economic sense of this term, it also invokes the notion of liberation as *moksha* — the process of our spiritual self-realization (Daniélou, 1993). The term, moksha, comes from the spiritual tradition of sanātana-dharma, which is commonly called Hinduism, but there are equivalent notions throughout the Dharmic and Taoic traditions. In the Abrahamic lineages, the notion typically gets confused with ideas about heaven being in some other time and place. However, with the notion of moksha, we find the possibility of psychospiritual disciplines that explore liberation in the here-and-now. We also find that moksha follows closely from *kāma* — the notion of *the desire of sensuality as a longing for the divine* (Daniélou, 1993). In short, once one accepts the notion of our embodiment as a conduit for the infinite flow of esoteric energies, one begins to appreciate these subtle energies as constituting the divinity of our humanity. The awareness of our experiential embodiment becomes a path of spiritual awakening.

If somatic psychology, in its focus on the subtle movements of our embodied experience, deconstructs the traditional dichotomy between the sacred and the secular, then we can anticipate its contribution to the development of what might be called secular spirituality. The notion of secular spirituality entails a non-theistic spiritual practice that operates without necessary reference to a personified deity or absolute

— it is spirituality quite unlike that enshrined by most organized religions. This spirituality is an *ethical-existential practice* which seems — almost universally — to have three coordinates, which I will mention briefly:

• The first coordinate of secular spiritual practice is that of Love. We surely mistake the nature of Love when we remain "in our heads" about it. Love is not a representation of something judged to be "good," as contrasted with something judged as "bad" (Barratt, 2004a, 2009b). It is not an idea about something, a decision to treat well the "object of love" (an act which is always at the expense of some other object that is treated less well). Love is not an attachment replete with obligations and expectations, nor even is it an affectionately beneficial state of affiliation. Rather, Love is something more like a vibration we can be attuned to — a process that transcends dualities and that transgresses the rules and regulations of everyday life. As a southern Athabaskan or Apachean proverb tells us, "Love is the real God of the entire world."

It is a platitude to suggest that one's capacity for Love is founded on one's capacity to find Love within oneself. Yet rarely have commentators investigated what this might mean — beyond the judgmental level of liking oneself, or being good to oneself. Love is at best limited if it operates under conditions of self-alienation and, in this sense, our potential to experience Love is intricately linked to our capacity to accept, appreciate and listen responsively to the voice of our embodied experience. Love is grounded in the experiences of embodiment, and the flow of subtle energies that run within, through and around our bodily being. In a different vein, Love is a communication or a circulation of energies between two or more persons, or between a person and the universe. It is the energetic power of transcendence.

• The second coordinate of secular spiritual practice is that of Freedom. In the western tradition, we tend to think in terms of freedom *from*, as in freedom from persecution, and of freedom *to*, as in freedom to express opinions (Fromm, 1990, 1994). However, this sort of conceptualization of freedom as a circumstantial condition actually limits our appreciation of the power of freedom within us (Barratt, 2004a, 2009b; Osho, 2004a, 2004b; Krishnamurti, 1975, 1996, 2007).

Francis of Assisi is reported to have proclaimed that "there are beautiful wild forces within us." If the practices of somatic psychology and bodymind therapy facilitate our awareness of this power within us, then what is discovered is surely *consciousness-as-movement* pervading the entire bodymind, and not just consciousness as a reflective operation occurring within the cerebrum? The conditions of alienation from the experience of our embodiment, the mind/body split, are an anathema to freedom. So too are all the blockages and obstructions to the free-flow of energies within us and to our awareness of them. In short, freedom begins with our potential for awareness of the multiplicity of communications within the experience of our own embodiment. Freedom begins with our experience of embodiment, and our ability to become aware of ourselves, to engage in this existential-spiritual process of self-realization (Barratt, 2009b). This freedom, the freedom to know the inner wildness of our embodied experience, dwells inherently in the bodymind's play of consciousness-as-movement. It is, in a profound sense, the ground and the prototype for freedom in every other sense of the term.

• The third coordinate of secular spiritual practice is that of Joy. The twentieth century Tantric mystic, Osho, once said "existence is made of the stuff called joy" (Osho, 2009). The aphorism is jarring given that there is so much suffering throughout the planet. Moreover, to our spiritually dissociated patterns of thinking, joy or happiness is the result of having something or of doing something; rather, than a condition of being, or more precisely of being-in-process, and living fully within one's embodiment. What Osho intends to communicate — along with every other great teacher from every other spiritual tradition — is that joy has little or nothing to do with money or material wealth, with political power or prestige, with social influence or gratifying interpersonal attachments, or with having a sense of moral superiority over other people who are viewed as inferior. These are the four great seductions of our egotism: wealth, power, popularity, and moral superiority (Barratt, 2009b). Rather, joy or authentic happiness inhere to the process of awareness, the process of spiritually awakening.

In this sense, it is embodied experience that grounds and makes possible our capacity to know joyfulness. In sickness or in health, it is our potential to connect with the embodied powers within us that makes possible our happiness — even in adverse circumstances. The processes of transcendence depend on our awareness of the processes of life itself. Reconnecting with our embodied experience, and our awareness of it, is the pathway for our joyful connection to the universe of the divine, because the subtle energies that circulate within us also flow through and around us, connecting our being-in-the-world with the entirety of the universe.

The teaching of every authentic spiritual tradition has recognized this interdependence between the interiority of our embodied experience and the divinity of the universe. We ended Chapter 10 with three quotations — from the Chandogya Upanishad, from Lao-Tzu, and Saraha — each suggesting the intricate interdependence between the interiority of our embodiment and the divinity of the universe. At least for the Dharmic and Taoic traditions, awareness of the interiority of our experiential embodiment leads to the awareness of the way in which this experience is cosmically embedded. Embodied awareness thus becomes a necessary and perhaps sufficient practice for the individual's authentic realization of his or her connection with the divine. While organized religions often view our bodily impulses as a moral liability, somatic psychology reopens the possibility of listening to the voice of our embodiment as an ethical and spiritual practice. In this respect, the discipline associates itself with mystical traditions that are both ancient and contemporaneous.

Although some mystical teachings, particularly within the Abrahamic tradition, have occasionally advocated excoriation and self-mortification of the body as a means to its transcendence, it would be a mistake to conclude that bodily awareness is no longer integral to spiritual practice in these situations. The ascetic repudiation of bodily pleasures as a spiritual practice does not necessarily imply that bodily awareness is devalued. It is clear from the writings of many esoteric practitioners that the contrary might be more correct; namely, that the intent of body-negating practices is precisely to enhance the acuity of awareness and to hone the experience of connection between the individual and the divine. Other mystical teachings, particularly within the Dharmic and Taoic traditions, have advocated sensual indulgence as a means to spiritual awareness. For example, the Mahabharata tells us that "pleasure is the basis of all the other aims of life," including that of spiritual transcendence. The importance of bodily pleasures in transgression and transcendence — the spiritual practices that dissolve our egotistic obstructions to spiritual awakening — is well articulated in the tradition of Tantric meditation. But contrary to some of its vulgarizations and mischaracterizations, Tantric methods are a rigorous and disciplined spiritual practice of attending to the movement of subtle energies within, through and around our embodiment, and most of these methods do not involve any sort of ostensible engagement with sensual pleasures (Barratt, 2006). Like the Tantric lineages, most mystical traditions follow Gautama Buddha's teachings of the *madhyamā-pratipad*, the "middle way," which suggests that neither ascetic self-mortification nor sensual licentiousness is necessarily helpful to spiritual practice, for the crucial issue is the cultivation of embodied awareness with the intent of spiritually awakening.

In almost all mystical traditions of spiritual practice, it is specifically indicated that awakening from the benumbed, deluded and alienated condition of our egotism — the separated condition of objectivistic consciousness — requires the cultivation of cosmic awareness, which begins with the awareness of embodied experience. In many such traditions, spiritual awakening is held to be either an orgasmic experience, or likened to such an experience. As was mentioned in Chapters 14 and 15, orgasming is understood here to be a full-bodied experience of intensely free-flowing energy rippling throughout our embodiment, and accompanied by the release of polypeptides (hypothalamic hormones and endorphins) as well as concomitantly altered states of consciousness. It is far more than the localized pelvic release which is the attenuated experience that many people think of as orgasm. Full-bodied or "total" orgasming, which can occur with considerable duration, involves a dedifferentiated experience of cosmic merging, in which partners no longer experience each other as separate, and in which the rush of subtle energies throughout the body is experienced as flowing freely into the energies of the entire cosmos. Such orgasming is a transcendent experience that operates spiritually to dissolve the fortifications of our egotism. It is, in a sense, the apotheosis of pleasure, which is understood by spiritual practitioners from every tradition to be a reflection of the infinite ecstasy that can be experienced by every individual as they become united with the universal or divine Being.

While this may seem like an idealized account of orgasmic experience, it serves to illustrate something crucial about the spiritual-ethical approach to an awareness of the bodymind's subtle energies. Except for the obstacles constructed by human egotism, these energies are seamlessly connected with the energies of the universe. This is the central tenet of mystical experience. The universe unfolds within the human body, just as the energies of our experiential embodiment issue into the energies of the entire universe. Somatic awareness flows into cosmic consciousness.

Eros is not only the inherent nature of our sexual and sensual corporeality; it is the nature of the universe itself. Cultivating an awareness of the interiority of our embodied experience thus becomes a spiritual practice with profound ethical and existential implications; the practitioners of somatic psychology and bodymind therapy need to acknowledge fully these implications. The cultivation of awareness of what is inside invariably issues into a transcendent awareness of all that appears to us to be outside of ourselves. This insight is found in the words of Novalis (the pseudonym of Georg von Hardenberg), the nineteenth century German author and romantic philosopher:

> There is only one temple in the world
> And that is the human body.
> Nothing is more sacred than that noble form.

The Bengali poet, Rabindranath Tagore, expressed this beautifully in the 69th stanza of his *Gitanjali*, written in 1912:

> The same stream of life that runs through my veins night and day runs through the world and dances in rhythmic measures. It is the same life that shoots in joy through the dust of earth in numberless blades of grass and breaks into tumultuous waves of leaves and flowers. It is the same life that is rocked in the ocean-cradle of birth and of death, in ebb and in flow. I feel my limbs are made glorious by the touch of this world of life ... the life-throb of ages dancing in my blood this very moment.

With slightly different wording, this is echoed in the words of Osho, the twentieth century Tantric mystic:

> The body is the visible soul, and the soul is the invisible body ... it is simply marvelous! And blessed are those who marvel. Begin the feeling of wonder with your own body, because that is the closest to you. The closest nature has approached you, the closest existence has come to you, is through your body. In your body is the water of the oceans, in your body is the fire of the stars and the suns, in your body is the air, your body is made of the earth. [Our body is our relationship with the existence of the universe, and God is] the experience that the whole universe is alive ... it has a heartbeat, and the moment you know that the universe has a heartbeat, you have discovered God.

The future of somatic psychology and bodymind therapy depends critically on our readiness to deconstruct the separation of science and spirituality that has characterized the modern western world. It will depend critically on our readiness to embrace the insight that cultivating the awareness of our embodied experience is an inherently spiritual practice.

17
The Future of Human Awareness

Dubliners, first published in 1914, James Joyce recounts the tragic story of Mr. Duffy, a bank cashier who "lived at a little distance from his body" (Joyce, 2006, p. 86). If you are familiar with the story you will recall that Mr. Duffy is befriended by a married woman who "urged him to let his nature open to the full" (p. 87). She longs for his touch and one evening "caught up his hand passionately and pressed it to her cheek" (p. 88). Mr. Duffy immediately rejects her advance, and terminates the relationship. Apparently heartbroken, she deteriorates into alcoholism and later suicides, crossing some railway tracks in front of an oncoming train. Mr. Duffy later learns of her fate in a newspaper obituary. He is initially revolted by the news of her addiction and her "commonplace vulgar death" (p. 90), but is eventually compelled to ask himself: "Why had he withheld life from her? Why had he sentenced her to death?" (p. 91).

In many respects — despite all our apparent indulgences — Mr. Duffy is actually the model of western man in the modern era (cf., Berman, 1989). Existing at a little distance from his bodily experience, he repels the sensual call of nature. Rather, he lives "outcast from life's feast" (p. 92). Whereas many of us express our repulsion by abusing and exploiting the "other," Mr. Duffy's repulsion is more notably expressed by withholding life from himself and from the woman who desires him — the woman who manifests life's longing for itself. Mr. Duffy is enslaved not only to his labors at the bank, and to the daily drudgery of his emotionally bleak routines, but also to his own pontifications about life, to his moralizing scruples and to his conceptual formulations about what ought to be. The notion of "living at a little distance" from one's own embodiment may initially seem almost humorous, partly because Joyce frames the implied notion of self-alienation in spatial terms. But the tragedy of western culture — for both men and women — is that we are socialized to live, like Mr. Duffy, "in our heads," which means temporally "out of sync" with the natural impulses of our own corporeal experience. We are not accustomed to live vibrantly in a dynamically estranged but creative dialogue with the voice of our experiential embodiment. Rather, we are acculturated to live with our body as we conceptualize it, as a machine that is alienated from the actuality of our desire and from the reality our bodily experience (Barratt, 1993).

In his 1992 book, Michael Murphy, co-founder of the Esalen Institute, invited us to consider the future of the body and the further evolution of human nature. In a somewhat similar manner, it seems appropriate to conclude this volume with a few comments on the possible future of embodied awareness. If indeed the limits of my language are the limits of my world (as has been variously articulated by diverse philosophers from Dilthey to Heidegger and Wittgenstein), then we need to ask what sort of language is the "language" of bodily awareness, and how is it articulated in relation to representational thinking and reflective consciousness.

Philosophers of language — consider here writers such as Charles Sanders Peirce and Ernst Cassirer to Roman Jakobson and even more contemporary formulations — have consistently shown that representational language is essentially metaphoric, in the general sense of this term (e.g., Cassirer, 1962). The language of representation depends on processes of condensation and displacement in which the meaning of a sign or symbol is entirely sustained by its relations with other signs and symbols (in principle, a minimum of three). One consequence of this structure is that the network of our representational system — within which our "reality" appears to be captured — is thus susceptible to the expression of duplicity, deceit, and ambivalence.

This leads to the paradox that the language of the mind, which I have been referring to as reflective consciousness, conceptual thinking or representationality, is able to trick itself (think here of all the ego's defense mechanisms documented by psychoanalysis), whereas the "language," in which the voice of our experiential embodiment expresses itself, perhaps cannot (cf., Schneider, 1999). Various assertions have been made that the "body never lies" or that it can be "read" unambiguously by those skilled in such a practice (Diamond, 1989; Frank, 2001; Kurtz & Prestera, 1984; Kushi, 2007; Ohashi, 1991; Olsen & McHose, 2004; Todd, 1980).

More cautiously, Totten suggests that the communicative capacity of the body has no equivalent to defense mechanisms such as projection (which are operations or transformations of representation). He raises the question how (not whether, but *how*) "does ambivalence express itself on a bodily level" (2003, p. 143). In my opinion, the question might be raised *whether* the "language" of the body can express ambivalence at all — if ambivalence is understood as the ability to communicate opposing valences simultaneously. Oppositionality is, after all, a rather distinctive feature of representational thinking, which defines *"a"* in terms of whatever is *"not-a"* (with *"not-a"* having to be defined with a third term). It thus opens us to the possibility of having emotional inclinations on both sides of the dichotomy (I feel *both* liking *and* disliking for so-and-so), or of enunciating sentiments on one side of the equation when the other side is actually more pertinent (my repulsion toward so-and-so wards off potential attraction). The ability to trick ourselves, to dissimulate, may be a device of the mind (reflective consciousness, conceptual thinking or representationality) that is not shared by bodily experience (the semiotics of our somatic being).

In Chapter 12, we briefly discussed the distinction between symbolic language — which permits the second-order operation of thinking about thinking that is the hallmark of reflective or secondary consciousness — versus the awareness to which somatic psychology refers that is sometimes called primary consciousness. As was indicated previously, our induction into symbolic language is implicated in our passage through Oedipal complexities. What is called primary consciousness, however, comprises a level of sensitivity and responsiveness to events — including affective dispositions — that may not even be available to conceptual formulation. Thus, the awareness furnished as primary consciousness cannot necessarily be translated into words (or can be translated into words, in an approximative manner, but with a significant loss of meaningfulness). Such awareness involves signs but not symbols (although some Jungians and other theorists would dispute this, arguing that symbols arise from within the somatic core of our being). Signs are communications that do not have the triadic structure of symbols (words or symbols are triadic in the sense that they only acquire meaning in relation to at least two other words or symbols).

If this distinction holds, then the meaningfulness of bodily awareness occurs in the domain of what might be called *somatic semiotics*. It is the bodymind's consciousness of its own corporeal experience — its receptivity to its own semiotics or system of signification and to all the energetic manifestations connected to it. This system might be described as nonsymbolic, presymbolic, preverbal, or preconscious (if "consciousness" is used in the limited sense of the symbolic realm of reflective consciousness). Awareness (the primary mode of somatic semiotics or whole-body consciousness) is attuned to the movement of subtle energies that flow within, through and around our embodiment. As indicated by Freud's notion of libidinality, this movement of the lifeforce cannot be adequately captured in the language of representational thinking (in Buddhist allegory, the finger that points to the moon should not be mistaken for the moon itself). The language of representational reflection finds the subtle movements of energies embraced by our embodiment to be thoroughly elusive.

To the extent that we are attuned to our corporeality, we enjoy the wisdom of knowing that there is a vastly expansive depth and breadth to our embodied experience. It is a depth and breadth that defies representation or narration. Moreover, if we are not so attuned, we still may have had moments in which the immediacy of awareness ruptures the hold of representationality over our experience. In innumerably varied ways, events such as near-death encounters, meditative processes, trips on lysergic acid diethylamide, ecstatic moments, and orgasmic dedifferentiation, all offer us — at the very least — a "breakthrough" in which we glimpse the limitation of living in the reality constituted by representations and narrations. Such breakthrough events open us to the depth and breadth of experiences that are otherwise than that which is established within the domain of ordinary thinking.

If the functional distinction between our awareness in the domain of somatic semiotics and reflective consciousness in the realm of representational thinking is indeed valid, then the key question for the future of somatic psychology concerns whatever we might be able to know about the connection between them.

On a practical level, this is a question with which Gendlin (1997) — to give just one notable example — has wrestled with considerable sophistication. His method of "focusing" directs the subject's attention to the inner cues of bodily experience, but then progressively attempts to refine the subject's ability to translate these feelings into verbal formulations. The latter aspect of this procedure may, or may not, be a mistake. There seems to be an assumption here — an assumption that is very much in the tradition of European phenomenology — that representationality is an open system, receptive to voices that come from beyond its own limitations. This ontological assumption implies that messages from the somatic semiotic domain can potentially be brought within, and realized within, the purview of reflective — verbal, symbolic — consciousness.

In this respect, there has always been an irrevocable divergence between phenomenological philosophy and psychodynamic practice (cf., Barratt, 1984, 1993). This divergence has crucial relevance to the contemporary deliberations and to the future of somatic psychology. The phenomenological tradition assumes that the experiences of primary consciousness are translatable, and therefore subsumable, into the language of reflective representation. By contrast, psychodynamics developed out of what has been dubbed the "school of suspicion," which advocates approaching the entire purview of reflective consciousness and conceptualization as principally an ideological system of "false consciousness" (Ricoeur, 1970, p. 33). This suspiciousness implies that the representational realm of reflective awareness is actually not so open. It does not merely await communications from the domain of somatic semiotics. Rather, the representational realm is structured in such a way as to suppress or repress our awareness of embodied experience (Barratt, 1993, 2005). As was discussed in Chapter 7, psychodynamics suggests that the interiority of meaningfulness is in this perpetual condition of movement and contradictoriness or conflict — such that any apparent resolution to such conflict is, at most, temporarily static or reified.

There is a radical divergence between the assumption that the representational system of reflective consciousness is open to, but has merely somehow *lost track of*, the messages of embodied experience, and the project of investigating the possibility that the narrations of representationality are ideologically structured so as to *block*, avoid or obfuscate, the communications of our embodiment. Phenomenological methods operate on the former assumption. Psychodynamic praxis adheres to the possibility that our attachment to the — illusory and perhaps even delusional — prerogatives of reflective consciousness, conceptual thinking or representationality, is precisely what perpetuates our alienation from the lifeforce presented to us through our embodied experience. There is a radical difference here in terms of the way in which the products of reflective consciousness are to be approached.

This is an issue around which the future of somatic psychology and bodymind therapies pivots. Are the practices of this discipline simply an expansion of the realm of representationality by the operations of reflective consciousness, or does the cultivation of somatic awareness actually deconstruct the ideological prerogatives of conceptual thinking? Of course, the answer to this question might be that both processes actually occur. However, such an answer is recursive in that it merely repositions the issue at an even deeper and more complex level: Is the praxis of somatic psychology an expansion of ordinary consciousness, or an act of ideology critique ... and if it is both, then when, where, and how, are these radically different directions operative?

To argue that — or to proceed as though — the practice of listening to the experiential voice of our embodiment merely expands the purview of reflective consciousness might be dangerous in a certain specific sense. Not only does such practice fail to address why it is that our reflective consciousness became alienated from our embodiment, it also might represent a serious collusion with the forces of suppression and repression. The expansion of reflective awareness would be a configurative labor of calculated reasoning about "what the body wants to say." Yet it is the mechanisms of configuration and calculation — including the representational devices of suppression, repression, and other "defenses" — that censored the voicing of our embodied experience in the first place. Just as Freud indicated that one cannot reason one's way into the desire of the unconscious that "reason" has itself repressed, one cannot truly express the voice of embodied experience in language, if the construction of that language is itself the cause of our alienation from that experience. Freud indirectly suggested that such insights on the part of reflective consciousness are as useful as the "distribution of menus in a time of famine" (1910, p. 123). In this context, a labor of bodymind therapy that merely listened, as if receptively, to whatever appears to be the voice of embodied experience would be, at best, of limited value to the patient's healing.

By contrast, if it is possible that representational consciousness and reflective reasoning are responsible for the maintenance of our alienation from the wisdom of embodied experience, the practice of listening to this wisdom requires "workplay" (a synergistic process that combines both work and play dimensions) that deconstructs the suppressive and repressive structuring of reflective consciousness (Barratt, 1993). To put this colloquially, the issue is no longer a refinement of conceptualizations about the body that are "in our heads," but rather a question of getting "out of our heads" and "into our bodies." Such a workplay that deconstructs the suppressive and repressive aspects of "living in our heads," would acknowledge that the communications of our embodiment can only be accessed by processes of indirection, and that healing necessarily involves a defigurative — playful — dimension.

In this respect, the point is not for reflective consciousness to treat the "language" of bodily experience as an "other" language to be translated into its own idiom — nor even as an "other" that is to be celebrated rather than subjugated. Rather, the intent is for reflective consciousness to undergo the deconstruction of its own suppressive and repressive aspects — a deprogramming of our blockages and obstructions to the free-flow of the lifeforce within, through and around us. By such dialectical and deconstructive method, the otherwise "language" of embodied wisdom comes into the workplay of our being-in-the-world.

This process is akin to the psychoanalytic discovery of free-associative discourse, in that its deconstructive momentum opens consciousness to the *otherwise* dimensions of our being that its structure suppresses and represses from itself. In this context, I understand free-associative discourse not as a means to an end (the end being greater scope of conceptual and narratological formulation), but rather as inherently healing (Barratt, 1993). Nothing about our experiential embodiment needs be treated as alien; yet healing requires that its wisdom be listened to as the voice of something strange, miraculous, and essential to our wellbeing and to our life itself.

Bodily awareness is the authentic ground of our being-in-the-world. Although we may become aware in the course of our empathic connectedness with the energies of the other and with the transpersonal energies of the natural universe, our potential for such awareness starts with, and is always founded on, our awareness of the experience of our corporeality. It is predictable that the eco-sensitive call for a return to an appreciative awareness of nature — as articulated with brilliance and sophistication by Erazim Kohák and others — will prove hollow, or of limited value, if it is not at least accompanied by a call for a return to an appreciation of embodied experience. Yet humanity urgently needs a revival of our capacity for the energetic interchange of energies that is fundamental to our empathy for whatever appears to us conceptually as "other." We urgently need to deconstruct the motif of mastery as domination and subjugation of the other (of people made subordinate and of the natural world). We urgently need to listen to the voice of our embodied experience as our personal path to the transpersonal and the transcendent.

The history of the westernized world is one of an escalating alienation from the processes of embodied awareness. This is, by the same token, an escalating infatuation with the motif of domination over the other, with the dichotomies of subject/object or mind/body, and with the material and technological achievements that result from "living in our head." We now need to return to a sense of belonging with our bodies — not to a program of conceptually evaluating them, improving them, or attempting to control their mechanics, but to listening to the wisdom that comes from their somatic semiotics. This is a revival of our knowledge of freedom and presence as the healing processes that honor the lifeforce itself. Once we dissolve its blockages and obstructions, our awareness of the wisdom of our embodiment opens us to an otherwise world from that which oppresses us today. It opens us to new possibilities for our human potential — culturally, politically, and spiritually. This then is the mandate of somatic psychology and bodymind therapy, and its potential for the prospective creation of profound change in our human condition cannot be overestimated.

Bibliography

Abram, D. (1997). *The spell of the sensuous: Perception and language in a more-than-human world*. New York, NY: Vintage.

Achterberg, J. (2002). *Imagery in healing: Shamanism and modern medicine*. Boston, MA: Shambhala.

——, Dossey, B. & Kolkmeier, L. (1994). *Rituals of healing: Using imagery for health and wellness*. New York, NY: Bantam.

Adorno, T. W. (1966). *Negative dialectics*. New York, NY: Seabury.

—— (1982). *Against epistemology: A metacritique*. London, UK: Blackwell.

—— (2001). *The culture industry*. New York, NY: Routledge.

Akhtar, S. (ed., 2006). *Interpersonal boundaries: Variations and violations*. Lanham, MD: Jason Aronson.

Adams, M. V. (2004). *The fantasy principle: Psychoanalysis of the imagination*. London, UK: Routledge.

Adler, A. (1998). *Understanding human nature* (trans. C. Brett). Center City, MN: Hazelden.

Adler, J. (2002). *Offering from the conscious body: The discipline of authentic movement*. Rochester, VT: Inner Traditions.

Ahsen, A. (1993). *Imagery paradigm: Imaginative consciousness in the experimental and clinical setting*. New York, NY: Brandon House.

Ajaya, S. (1983). *Psychotherapy east and west: A unifying paradigm*. Honesdale, PA: Himalayan Institute Press.

—— (2008). *Healing the whole person: Applications of Yoga psychotherapy*. Honesdale, PA: Himalayan Institute Press.

Albretch, G.L., Fitzpatrick, R. & Scrimshaw, S. (2000). *Handbook of social studies in health and medicine*. London, UK: Sage.

Alexander, F. M. & Barlow, W. (2001). *The use of the self*. Ukiah, CA: Orion Publishing.

Alexander, G. (1981). *Eutony: The holistic discovery of the total person*. New York: Felix Morrow.

Allport, S. (2001). *Explorations of the black box: The search for the cellular basis of memory*. Bloomington, IN: AuthorHouse.

Almaas, A. H. (2000a). *Diamond heart* (four volumes). Boston, MA: Shambhala.

—— (2000b). *Pearl beyond price: Integration of personality into being, an object-relations approach*. Boston, MA: Shambhala.

—— (2000c). *The void: Inner spaciousness and ego structure*. Boston, MA: Shambhala.

—— (2004). *Inner journey home: The soul's realization of the unity of reality*. Boston, MA: Shambhala.

—— (2008). *The unfolding now: Realizing your true nature through the practice of presence*. Boston, MA: Shambhala.

Almond, G. A., Appleby, R. S. & Sivan, E. (2003). *Strong religions: The rise of fundamentalisms around the world*. Chicago, IL: University of Chicago Press.

Alschuler, L. (2006). *The psychopolitics of liberation: Political consciousness from a Jungian perspective*. New York, NY: Palgrave Macmillan.

Altman, N. (1995). *The analyst in the inner city: Race, class, and culture through a psychoanalytic lens.* Hillsdale, NJ: Analytic Press.

Anderson, F. S. (2008). *Bodies in treatment: The unspoken dimension.* New York, NY: Analytic Press.

Anderson, J. (1993). *Ballet and modern dance: A concise history* (2nd ed.). Princeton, NJ: Princeton Book Company.

Anderson, W. T. (1983). *The upstart spring: Esalen and the human potential movement — the first twenty years.* Reading, MA: Addison Wesley.

Anzieu, D. (1989). *The skin ego: A psychoanalytic approach to the self.* New Haven, CT: Yale University Press.

—— (ed., 1990). *Psychic envelopes.* London, UK: Karnac Books.

—— (1995). *Le moi-peau* (2nd ed.). Paris, France: Dunod.

Apfelbaum, A. (2003). *Thai massage: Sacred body work.* Knoxville, TN: Avery.

Aposhyan, S. (1999). *Natural intelligence: Body-mind integration and human development.* Baltimore, MD: Williams & Wilkins.

—— (2004). *Body-mind psychotherapy: Principles, techniques, and practical applications.* New York, NY: W. W. Norton.

Appadurai, A. (ed., 2001). *Globalization.* Durham, NC: Duke University Press.

Arnold, D. (2008). *Buddhists, brahmins and belief: Epistemology in South Asian philosophy.* New York, NY: Columbia University Press.

Aron, L. & Harris, A. (eds, 1993). *The legacy of Sandor Ferenczi.* Hillsdale, NJ: Analytic Press.

Ashcroft, B. (ed., 2008). *Post-colonial studies: The key concepts.* New York, NY: Routledge.

Atreya (1996). *Prana: The secret of Yogic healing.* Santa Rosa, CA: Atrium Publishing Group.

Au, S. (2002). *Ballet and modern dance.* London, UK: Thames Hudson.

Ayto, J. (2001). *The origin of words: The histories of over 8,000 words explained.* London, UK: Bloomsbury Publishing.

Baer, H., Singer, M. & Susser, I. (2003). *Medical anthropology and the world system.* Westport, CT: Praeger.

Baker, E. F. (1967). *Man in the trap: The causes of blocked sexual energy.* London, UK: Macmillan.

—— & Reich, W. (1955). *Medical orgonomy.* Rangeley, ME: American Association for Medical Orgonomy.

Baldwin, J. (1961). *Nobody knows my name.* New York, NY: Dell.

Balint, M. (1987). *Thrills and regressions.* London, UK: Karnac Books.

—— (1995). *Primary Love and Psychoanalytic Technique.* New York: St. Martins Press.

Bankart, C. P. (1997). *Talking cures: A history of western and eastern psychotherapies.* Pacific Grove, CA: Brooks/Cole.

Barnard, A. (2002). *Encyclopedia of social and cultural anthropology.* New York, NY: Routledge.

Barratt, B. B. (1984). *Psychic reality and psychoanalytic knowing.* Hillsdale, NJ: Analytic Press.

—— (1985). Psychoanalysis as critique of ideology. *Psychoanalytic Inquiry, 5*: 437–470.

—— (1993). *Psychoanalysis and the postmodern impulse: Knowing and being since Freud's psychology.* Baltimore, MD: Johns Hopkins University Press.

—— (1999). Cracks: On castration, death and laughter. In: J. Barron (ed.), *Humor and psyche*. New Haven, CT: Yale University Press, pp. 57–67.

—— (2004a). *The way of the bodyprayerpath*. Philadelphia, PA: Xlibris/Random House.

—— (2004b). Desire and death in the constitution of I-ness. In: J. Reppen, J. Tucker & M. A. Schulman (eds), *Way beyond Freud: Postmodern psychoanalysis observed*. London, UK: Open Gate Press, pp. 264–279.

—— (2005). *Sexual health and erotic freedom*. Philadelphia, PA: Xlibris/Random House.

—— (2006). *What is tantric practice?* Philadelphia, PA: Xlibris/Random House.

—— (2009a). Ganesha's lessons for psychoanalysis: Notes on fathers and sons, sexuality and death. *Psychoanalysis, Culture and Society*, *14*: 317–336.

—— (2009b). *Liberating Eros*. Philadelphia, PA: Xlibris/Random House.

Bartenieff, I. (1980). *Body movement: Coping with the environment*. New York, NY: Routledge.

Barthes, R. (1979). *A lover's discourse: Fragments*. New York, NY: Hill & Wang, 1979.

Baruš, I. (2003). *Alterations of consciousness: An empirical analysis for social scientists*. Washington, DC: American Psychological Association.

Bataille, G. (1988). *Inner experience*. Albany, NY: State University of New York Press.

Baum, W. M. (2005). *Understanding behaviorism: Behavior, culture and evolution*. London, UK: Blackwell.

Baylis, J., Smith, S. & Owens, P. (2008). *The globalization of world politics* (4th ed.). New York, NY: Oxford University Press.

Becker, E. (1998). *The denial of death*. Boston, MA: Free Press.

Behar-Horenstein, L. S. & Ganet-Sigel, J. (1999). *The art and practice of dance/movement therapy*. Boston, MA: Pearson Custom Publishing.

Benhabib, S., Butler, J., Cornell, D. & Fraser, N. (1995). *Feminist contentions: A philosophical exchange*. New York, NY: Routledge.

Beresford-Cooke, C. (2003). *Shaitsu theory and practice: A comprehensive text for the student and professional*. Edinburgh, UK: Churchill Livingstone.

Bergson, H. (2007a). *The creative mind: An introduction to metaphysics*. New York, NY: Dover Publications.

—— (2007b). *Matter and memory*. New York, NY: Cosimo Classics.

—— (2008). *Time and free will: An essay on the immediate data of consciousness*. New York, NY: Cosimo Classics.

—— (2009). *Creative evolution*. London, UK: Palgrave Macmillan.

Berman, M. (1989). *Coming to our senses: Body and spirit in the hidden history of the west*. New York, NY: Simon & Schuster.

Bernard, A., Steinmuller, W. & Stricker, U. (2006). *Ideokinesis: A creative approach to human movement and body alignment*. Berkeley, CA: North Atlantic Books.

Bettelheim, B. (1983). *Freud and man's soul: An important re-interpretation of Freudian theory*. New York, NY: Vintage.

Bhatt, S. R. & Mehrotra, A. (2000). *Buddhist epistemology*. Westport, CT: Greenwood Press.

Bischoff, E. (1985). *Kabbala: An introduction to Jewish mysticism and its secret doctrine*. Newburyport, MA: Weiser.

Blacking, J. (1977). *The anthropology of the body*. London, UK: Academic Press.

Blakeslee, S. & Blakeslee, M. (2007). *The body has a mind of its own.* New York, NY: Random House.

Blechner, M. J. (2009). *Sex changes: Transformations in society and psychoanalysis.* New York, NY: Routledge.

Bloom, K. (2006). *The embodied self: Movement and psychoanalysis.* London, UK: Karnac.

Boadella, D. (1986). *Wilhelm Reich: The evolution of his work.* New York, NY: Arkana.

—— (1987). *Lifestreams: An introduction to biosynthesis.* London, UK: Routledge Kegan & Paul.

—— (ed., 1991). *In the wake of Reich* (2nd ed.). London, UK: Coventure.

Bolen, J. S. (2005). *The Tao of psychology: Synchronicity and the self.* New York, NY: HarperOne.

Bosnak, R. (2007). *Embodiment: Creative imagination in medicine, art and travel.* London, UK: Routledge.

Bowler, G. (2004). *Touching and communicating with energy* (3rd ed.). Phoenix, AZ: Energy Medicine Publishing.

Boyeson, G. (1994). *Über den Körper die Seele heilen: Biodynamische Psychologie und Psychotherapie.* Munich, Germany: Kösel.

—— & Boyeson, M. L. (1987). *Biodynamik des Lebens: Die Gerda Boyesen Methode: Grundlage der biodynamischen Psychologie.* Essen, Germany.

Braud, W. & Anderson, R. (eds, 1998). *Transpersonal research methods for the social sciences: Honoring human experience.* London, UK: Sage Publications.

Breaux, C. (1989). *Journey into consciousness: The chakras, tantra and Jungian psychology.* York Beach, ME: Nicolas-Hays.

Brentano, F. C. (1995a). *Psychology from an empirical standpoint* (2nd ed.). London, UK: Routledge.

—— (1995b). *Descriptive psychology.* London, UK: Routledge.

—— (2001). *The origin of our knowledge of right and wrong: With a biographical note.* Boston, MA: Adamant Media Corporation.

Britton, P. (2005). *The art of sex coaching.* New York, NY: W. W. Norton.

Brock, A. (2006). *Internationalizing the history of psychology.* New York, NY: New York University Press.

Bronfenbrenner, U. (2006). *The ecology of human development: Experiments by nature and design.* Cambridge, MA: Harvard University Press.

Broswimmer, F. (2002). *Ecocide: A short history of mass extinction of species.* London, UK: Pluto Press.

Brown, C. S. & Toadvine, T. (eds, 2003). *Eco-phenomenology: Back to the earth itself.* Albany, NY: State University of New York Press.

Brown, D. (2007). *Sexual surrogate partner therapy: Exploding the myths, exploring the mystery, explaining the therapy.* Luton, UK: JNB Publications.

Brown, P. (1988). *The body and society: Men, women, and sexual renunciation in early Christianity.* New York, NY: Columbia University Press.

Brown, Ph. (1974). *Toward a Marxist psychology.* New York, NY: Harper & Row.

Buber, M. (1971). *I and thou.* Boston, MA: Free Press.

—— (2002). *Between man and man* (2nd ed.). New York, NY: Routledge.

Buckley, K. W. (1989). *Mechanical man: John Broadus Watson and the beginnings of behaviorism.* New York, NY: Guilford Press.

Bugental, J. F. T. (1964). The third force in psychology. *Journal of Humanistic Psychology, 4*: 19–25.

Bugental, J. F. T. (ed., 1967). *Challenges of humanistic psychology.* New York NY: McGraw-Hill.

—— (1986). *The search for existential identity.* San Francisco, CA: Jossey-Bass.

Bullough, V. (1994). *Human sexuality: An encyclopedia.* New York, NY: Routledge.

Burmeister, M. (1997). *The touch of healing: Energizing the body, mind, and spirit with jin shin jyutsu.* New York, NY: Bantam.

Cabeza, R. & Kingstone, A. (2006). *Handbook of functional neuroimaging of cognition* (2nd ed.). Cambridge, MA: M.I.T. Press.

Calais-Germain, B. (2006). *Anatomy of breathing.* Seattle, WA: Eastland Press.

—— (2007). *Anatomy of movement.* Seattle, WA: Eastland Press.

Caldwell, C. (1996). *Getting our bodies back: Recovery, healing, and transformation through body-centered psychotherapy.* Boston, MA: Shambhala.

—— (ed., 1997). *Getting in touch: The guide to new body-centered therapies.* Wheaton, IL: Quest.

Calvert, R. N. (2002). *The history of massage: An illustrated survey from around the world.* Rochester VT: Healing Arts Press.

Camic, P. M., Rhodes, J. E. & Yardley, L. (2003). *Qualitative research in psychology: Expanding perspectives in methodology and design.* Washington, DC: American Psychological Association.

Campbell, J. (ed., 1985). *Spiritual disciplines: Papers from the Eranos yearbooks.* Princeton, NJ: Princeton University Press.

—— (1991). *The masks of God* (Four volumes). London, UK: Penguin.

Carlisi, A. P. (2007). *The only way out is in: A modern day yogi's commentary on the synergy of ashtanga yoga, âyurveda and tantra.* Hanalei, HI: Dream Weavers International.

Carter, C. S., Ahnert, K., Grossmann, L. E., Hrdy, S. B., Lamb, M. E., Porges, S. W. & Sachser, N. (2006). *Attachment and bonding: A new synthesis.* Cambridge, MA: M.I.T. Press.

Cassirer, E. (1962). *An essay on man: An introduction to a philosophy of human culture.* New Haven CT: Yale University Press.

Castaneda, C. (2008). *The teachings of Don Juan: A Yaqui way of knowledge.* Berkeley, CA: University of California Press.

Cater, N. (2005). Body and soul: Honoring Marion Woodman. *Journal of Archetype and Culture.* New Orleans, LA: Spring Journal.

Celenza, A. (2007). *Sexual boundary violations: Therapeutic, supervisory, and academic contexts.* Lanham, MD: Jason Aronson.

Certeau, M. de (2002). *The practice of everyday life.* Berkeley, CA: University of California Press.

Césaire, A. (2001). *Discourse on colonialism.* New York, NY: Monthly Review Press.

Chai, R. M. R. (ed., 2005). *Na mo'olelo lomilomi: The traditions of Hawaiian massage and healing.* Honolulu, HI: Bishop Museum Press.

—— (2007). *Hawaiian massage lomilomi: Sacred touch of aloha.* Kailua, HI: Hawaiian Insights.

Chalmers, D. J. (1997). *The conscious mind: In search of a fundamental theory.* New York, NY: Oxford University Press.

—— (ed., 2002). *Philosophy of mind: Classical and contemporary readings.* New York, NY: Oxford University Press.

——, Manley, D. & Wasserman, R. (eds, 2009). *Metaphysics: New essays on the foundations of ontology*. New York, NY: Oxford University Press.

Chia, M. & Arava, D. A. (1996). *The multi-orgasmic man*. San Francisco, CA: HarperSanFrancisco.

Chodorow, J. (1991). *Dance therapy and depth psychology: The moving imagination*. London, UK: Routledge.

Chomsky, N. (1959). "A Review of B. F. Skinner's Verbal Behavior". *Language*, 35: 26–58.

Chopra, D. (1989). *Quantum healing: Exploring the frontiers of mind/body medicine*. New York, NY: Bantam Books.

Chow, K. T. (2004). *Thai yoga massage: A dynamic therapy for physical well-being and spiritual energy*. Rochester, VT: Healing Arts Press.

Cixous, H. & Clément, C. (1986). *The newly born woman*. Minneapolis, MN: University of Minnesota Press.

Clark, A. (1998). *Being there: Putting brain, body, and world together again*. Cambridge, MA: M.I.T. Press.

—— (2008). *Supersizing the mind: Embodiment, action, and cognitive extension*. New York, NY: Oxford University Press.

Clayton, P. (2006). *Mind and emergence: From quantum to consciousness*. New York, NY: Oxford University Press.

Coakley, S. (ed., 2000). *Religion and the body*. Cambridge, UK: Cambridge University Press.

Cockerham, W. C. (2006). *Medical sociology* (10th ed.). Upper Saddle River, NJ: Prentice Hall.

Cohen, B. B. (2008). *Sensing, feeling, and action: The experiential anatomy of body-mind centering* (2nd ed.). Toronto, Canada: Contact Editions.

Comer, R. J. (2009). *Abnormal psychology* (7th ed.). New York, NY: W. H. Freeman.

Connor, L. H. & Samuel, G. (eds, 2001). *Healing powers and modernity: Traditional medicine, shamanism, and science in Asian societies*. Westport, CT: Bergin & Garvey.

Conrad, E. (2007). *Life on land: The story of Continuum*. Berkeley, CA: North Atlantic Books.

Conrad, P. (2008). *The sociology of health and illness: Critical perspectives*. New York, NY: Worth Publishers.

Cooper, D. G. (ed., 1967). *Psychiatry and anti-psychiatry*. Boulder, CO: Paladin Press.

—— (ed., 1968). *The dialectics of liberation*. London, UK: Penguin.

—— (1978). *The language of madness*. London, UK: Penguin.

Cornell, A. W. (2005). *The radical acceptance of everything: Living a focusing life*. Berkeley, CA: Calluna Press. [See also, Weiser, 1996]

Costello, M. S. (2006). *Imagination, illness and injury: Jungian psychology and the somatic dimensions of perception*. New York, NY: Routledge.

Coster, G. (2003). *Yoga and western psychology*. Whitefish, MT: Kessinger Publishing.

Courtois, C. A. & Ford, J. D. (eds, 2009). *Treating complex traumatic stress disorders*. New York, NY: Guilford Press.

Cowan, G. A., Pines, D. & Meltzer D. (1999). *Complexity: Metaphors, models, and reality*. Boulder, CO: Westview Press.

Cowan, T. (1993). *Fire in the head: Shamanism and the celtic spirit*. New York, NY: HarperOne.

Cozolino, L. (2002). *The neuroscience of psychotherapy: Building and rebuilding the human brain.* New York, NY: W. W. Norton.

—— (2006). *The neuroscience of human relationships.* New York, NY: W. W. Norton.

Cramer, P. (2006). *Protecting the self: Defense mechanisms in action.* New York, NY: Guilford Press.

Cravioto, M. A. S. (2007). *Awakening exercises for students of the fourth way: From the works of Gurdjieff and Ouspensky.* Boulder, CO: G. L. Design.

Crawford, I. (1990). *A guide to the mysteries.* London, UK: Lucis Press.

Crick, F. (1995). *Astonishing hypothesis: The scientific search for the soul.* New York, NY: Scribner.

Critchley, S. (2004). *Very little ... almost nothing: Death, philosophy and literature.* London, UK: Routledge.

Cross, J. R. (2006). *Healing with the chakra energy system.* Berkeley, CA: North Atlantic Books.

Csikszentmihalyi, M. (1999). *Flow: The psychology of optimal experience.* New York, NY: Harper Perennial.

Csordas, T. J. (ed., 1994). *Embodiment and experience: The existential ground of culture and self.* London, UK: Cambridge University Press.

Cushman, P. (1996). *Constructing the self, constructing America: A cultural history of psychotherapy.* Cambridge, MA: Da Capo Press.

D'Aquili, E. & Newberg, A. B. (1999). *Mystical mind: Probing the biology of religious experience.* Minneapolis, MN: Augsburg Fortress Publishers.

Dalai Lama, H. H. (1999). *Ethics for the new millennium.* New York, NY: Riverhead Books.

Dale, C. (2009). *The subtle body: An encyclopedia of your energetic anatomy.* Boulder, CO: Sounds True.

Damasio, A. (2000). *The feeling of what happens: Body and emotion in the making of consciousness.* Orlando, FL: Harcourt Press.

—— (2003). *Looking for Spinoza: Joy, sorrow, and the feeling brain.* Boston, MA: Mariner.

—— (2005). *Descartes' error: Emotion, reason, and the human brain.* New York, NY: Penguin.

Daniélou, A. (1993). *Virtue, success, pleasure, liberation: The four aims of life in the tradition of ancient India.* Rochester, VT: Inner Traditions.

Danto, E. (2005). *Freud's free clinics: Psychoanalysis and social justice, 1918–1938.* New York, NY: Columbia University Press.

Daruna, J. H. (2004). *Introduction to psychoneuroimmunology.* San Diego, CA: Elsevier Academic Press.

Davidson, J. (2004). *Subtle energy.* New York, NY: Random House.

Davidson, R. J. & Harrington, A. (eds, 2001). *Visions of compassion: Western scientists and Tibetan Buddhists examine human nature.* Oxford, UK: Oxford University Press.

Davies, E. (2006). *Beyond dance: Laban's legacy of movement analysis.* New York, NY: Routledge.

Deikman, A. J. (1982). *The observing self: Mysticism and psychotherapy.* Boston, MA: Beacon Press.

Deleuze, G. (1990). *Bergsonism.* San Francisco, CA: Zone Books.

—— (1995). *Difference and repetition.* New York, NY: Columbia University Press.

—— (2005). *The logic of sense.* London, UK: Continuum International.

—— & Guattari, F. (2004). *Thousand plateaus*. London, UK: Continuum International.

—— & —— (2009). *Anti-Oedipus: Capitalism and schizophrenia*. London, UK: Penguin.

Dennett, D. C. (1992). *Consciousness explained*. New York: Back Bay Books.

—— (1997). *Kinds of minds: Toward an understanding of consciousness*. New York, NY: Basic Books.

—— (2007). *Breaking the spell: Religion as a natural phenomenon*. New York, NY: Penguin.

Derrida, J. (1985). *Writing and difference* (trans. A. Bass). London, UK: Routledge.

—— (1989). *The postcard*. Chicago, IL: University of Chicago Press.

—— (1996). *The gift of death* (trans. D. Wills). Chicago, IL: University of Chicago Press.

—— (1999). *Resistances of psychoanalysis*. Palo Alto, CA: Stanford University Press.

—— (2001). *On cosmopolitanism and forgiveness*. New York, NY: Routledge.

Devereaux, G. (1980). *Basic problems of ethnopsychiatry*. Chicago, IL: University of Chicago Press.

Diamond, J. (1978). *Behavioural kinesiology and the autonomic nervous system*. Valley Cottage, NY: Archaeus Press.

—— (1985). *Life energy*. St. Paul, MN: Paragon House.

—— (1989). *The body doesn't lie: Unlock the power of your natural energy*. New York: Grand Central Publishing.

Doi, T. (1971). *The anatomy of dependence*. New York, NY: Kodansha International.

Douglas-Klotz, N. (2005). *The Sufi book of life: 99 pathways of the heart for the modern dervish*. London, UK: Penguin.

Dreyfus, H. L. (1972). *What computers can't do: The limits of artificial intelligence*. New York, NY: HarperCollins.

—— (1992). *What computers still can't do: A critique of artificial reason*. Cambridge, MA: MIT Press.

—— & Dreyfus, S. E. (1992). *Mind over machine*. Boston, MA: Free Press.

DuBois, T. A. (2009). *An introduction to shamanism*. London, UK: Cambridge University Press.

Dussel, E. (1985). *Philosophy of liberation*. Eugene, OR: Wipf & Stock.

Edelman, G. M. (1990). *Remembered present: A biological theory of consciousness*. New York, NY: Basic Books.

—— (1993). *Brilliant air, brilliant fire: On the matter of the mind*. New York, NY: Basic Books.

—— (2005). *Wider than the sky: The phenomenal gift of consciousness*. New Haven, CT: Yale University Press.

—— (2007). *Second nature: Brain science and human knowledge*. New Haven, CT: Yale University Press.

—— & Tononi, G. (2001). *A universe of consciousness: How matter becomes imagination*. New York, NY: Basic Books.

Eden, D. & Feinstein, D. (1999). *Energy medicine*. New York, NY: Tarcher.

Efron, A. (1985). *The sexual body: An interdisciplinary perspective*. Special issue of *Journal of Mind and Behavior*, 6 (1 & 2).

Ehrenberg, D. B. (1992). *The intimate edge: Extending the reach of psychoanalytic interaction*. New York, NY: W. W. Norton.

Eliade, M. (2004). *Shamanism: Archaic techniques of ecstasy*. Princeton, NJ: Princeton University Press.

Elias, N. (2000). *The civilizing process: Sociogenetic and psychogenetic investigations* nd ed.). Oxford, UK: Wiley-Blackwell.

Ember, C. (ed., 2004). *Encyclopedia of medical anthropology: Health and illness in the world's cultures*. New York, NY: Springer.

Epstein, M. (1995). *Thoughts without a thinker: Psychotherapy from a Buddhist perspective*. New York, NY: Basic Books.

—— (2008). *Psychotherapy without a self: A Buddhist perspective*. New Haven, CT: Yale University Press.

Erikson, E. H. (1995). *Childhood and society*. New York, NY: Vintage.

Ernst, C. W. (1997). *The Shambhala guide to Sufism: An essential introduction to the philosophy and practice of the mystical tradition of Islam*. Boston, MA: Shambhala.

Ernst, W. & Harris, B. (eds, 1999). *Race, science and medicine 1700–1960*. London, UK: Routledge.

Esman, A. (ed., 1990). *Essential papers on transference*. New York, NY: New York University Press.

Fay, B. (1996). *Contemporary philosophy of social science: A multicultural approach*. Cambridge, MA: Blackwell.

Fanon, F. (2005). *The wretched of the earth*. New York, NY: Grove Press.

—— (2008). *Black skin, white masks*. New York, NY: Grove Press.

Featherstone, M., Hepworth, M. & Turner, B. S. (1991). *The body: Social process and cultural theory*. London, UK: Sage Publications.

Feher, M., Nadaff, R. & Tazi, N. (eds, 1989). *Fragments for a history of the human body* (3 vols). New York, NY: Zone Books.

Feinstein, D., Eden, D. & Craig, G. (2005). *The promise of energy psychology: Revolutionary tools for dramatic personal change*. New York, NY: Tarcher/ Penguin.

Feldenkrais, M. (1981). *The elusive obvious: Or basic Feldenkrais*. Capitola, CA: Meta Publications.

—— (1991). *Awareness through movement: Easy-to-do exercises to improve your posture, vision, imagination, and personal awareness*. New York: NY: HarperOne.

—— (2002). *The potent self: A study of spontaneity and compulsion*. Mumbai, India: Frog Books.

—— (2005). *Body and mature behavior: A study of anxiety, sex, gravitation, and learning*. Mumbai, India: Frog Books.

Ferenczi, S. (1989). *Thalassa: A theory of genitality*. London, UK: Karnac Books.

—— (2008). *Further contributions to the theory and technique of psychoanalysis*. Leesburg, FL: Ford Press.

—— & Rank, O. (1956). *Sex in psychoanalysis & the development of psychoanalysis*. New York, NY: Dover Publications.

Ferrer, J. N. (2002). *Revisioning transpersonal theory: A participatory vision of human spirituality*. Albany, NY: State University of New York Press.

Feuerstein, G. (1989). *The Yoga-Sûtra of Patanjali: A new translation and commentary*. Rochester, VT: Inner Traditions.

—— (1998a). *The Yoga tradition: Its history, literature, philosophy and practice*. Prescott, AZ: Hohm Press.

—— (1998b). *Tantra: The path of ecstasy*. Boston, MA: Shambhala Press.

—— (2000). *The Shambhala encyclopedia of Yoga.* Boston, MA: Shambhala Press.

Fields, G. P. (2001). *Religious therapeutics: Body and health in Yoga, Āyurveda, and Tantra.* Albany, NY: State University of New York.

Fields, R. (1992). *How the swans came to the lake: A narrative history of Buddhism in America* (3rd ed.). Boston, MA: Shambhala Press.

Finke, R. A., Ward, T. B. & Smith, S. M. (1996). *Creative cognition: Research, theory, and applications.* Cambridge, MA: M.I.T. Press.

Fisher, A. (2002). *Radical ecopsychology: Psychology in the service of life.* Albany, NY: State University of New York Press.

Forrester, J. (1991). *The seductions of psychoanalysis: Freud, Lacan, and Derrida.* London, UK: Cambridge University Press.

Foucault, M. (1966). *The order of things: An archaeology of human sciences.* New York, NY: Vintage, 1994.

—— (1969). *The archaeology of knowledge* (trans. A. M. Sheridan Smith). New York, NY: Pantheon, 1972.

—— (1988). *Madness and civilization: A history of insanity in the age of reason.* New York, NY: Vintage Books.

—— (1988–1990). *The history of sexuality* (3 vols). New York, NY: Vintage.

Fox, D. R., Prilleltensky, I. & Austin, S. (eds, 2009). *Critical psychology: An introduction* (2nd ed.). London, UK: Sage.

Fox, M. (2002). *Creativity: Where the divine and the human meet.* New York, NY: Tarcher.

Francfort, H. & Hamayon, R. (eds, 2001). *The concept of shamanism: Uses and abuses.* Budapest, Hungary: Akademiai Kiado.

Frank, R. (2001). *Body of awareness: A somatic and developmental approach to psychotherapy.* Cambridge, MA: Gestalt Press.

Frankl, V. (1937). Zur geistigen problematik der psychotherapies. *Zentralblatt für Psychotherapie, 10*: 33–45.

—— (1984). *Man's search for meaning: An introduction to logotherapy.* New York, NY: Simon & Schuster.

Frantzis, B. (2008). *The chi revolution: Harnessing the healing power of your life force.* New York. NY: Blue Snake Books.

Fraser, M. (2005). *The body: A reader.* New York, NY: Routledge.

Frawley, D. (1997). *Ayurveda and the mind: The healing of consciousness.* Twin Lakes, WI: Lotus Press.

—— (2000). *Ayurvedic healing: A comprehensive guide* (2nd ed.). Twin Lakes, WI: Lotus Press.

Freedheim, D. K. (ed., 1992). *History of psychotherapy: A century of change.* Washington, DC: American Psychological Association.

Freud, A. (1936). *The Writings of Anna Freud* (4 vols). New York, NY: International Universities Press, 1966.

Freud, S. (1900). Die Traumdeutung. *Gesammelte Werke, 2–3*: 1–162. [Trans.: The Interpretation of Dreams. *Standard Edition, 4–5*: 1–627.]

—— (1910). Über "Wilde" Psychoanalyse. *Gesammelte Werke, 8*: 118–125. [Trans.: "Wild" psychoanalysis. *Standard Edition, 11*: 221–227.]

—— (1916–1917). Vorlesungen zur Einführung in die Psychoanalyse. *Gesammelte Werke, 11.* [Trans.: Introductory Lectures on Psychoanalysis. *Standard Edition, 15–16.*]

—— (1920). Jenseits des Lustprinzips. *Gesammelte Werke, 13*: 133–169. [Trans.: Beyond the Pleasure Principle. *Standard Edition, 17*: 7–64.]

—— (1923a). Das Ich und das Es. *Gesammelte Werke, 13*: 237–289. [Trans.: The Ego and the Id. *Standard Edition, 19*: 12–66.]

—— (1923b). 'Psychoanalyse' und 'Libidotheorie.' *Gesammelte Werke, 13*: 211–233. [Trans.: Two Encyclopedia Articles. *Standard Edition, 18*: 235–259.]

—— (1925). Die Verneinung. *Gesammelte Werke, 14*: 11–15. [Trans.: Negation. *Standard Edition, 19*: 235–239.]

—— (1926a). Die Frage der Laienanalyse. *Gesammelte Werke, 14*: 209–286. [Trans.: The Question of Lay Analysis. *Standard Edition, 20*: 183–250.]

—— (1926b). *Hemmung, Symptom und Angst. Gesammelte Werke, 14*: 113–205. [Trans.: Inhibitions, Symptoms and Anxiety. *Standard Edition, 20*: 87–172.]

—— (1933). Neue Folge der Vorlesungen zur Einführung in die Psychoanalyse. *Gesammelte Werke, 15*. [Trans.: New Introductory Lectures on Psychoanalysis. *Standard Edition, 22*: 5–182.]

—— & Jung, C. G. (1994). *The Freud/Jung Letters* (ed., W. McGuire; trans., R. F. C. Hull). Princeton, NJ: Princeton University Press.

Frick, C. A. (1994). *The cognitive turn: The interdisciplinary story of thought in western culture*. Lanham, MD: University Press of America.

Fritz, S. & Daholsky, K. M. (2000). *Mosby's basic science for soft tissue and movement therapies*. Edinburgh, UK: Churchill Livingstone.

—— & Grosenbach, J. (2008). *Mosby's essential sciences for therapeutic massage: Anatomy, physiology, biomechanics and pathology* (3rd ed.). New York: Mosby.

Fromm, E. (1994). *Man for himself: An inquiry into the psychology of ethics*. New York, NY: Holt Paperbacks.

—— (1994). *Escape from freedom*. New York, NY: Holt Paperbacks.

—— (2005). *To have or to be?* New York, NY: Continuum.

Fuller, S., De Mey, M., Shinn, T. & Woolgar, S. (eds, 1989). *The cognitive turn: Sociological and psychological perspectives on science*. New York, NY: Springer.

Furness, J. B. (2006). *The enteric nervous system*. Oxford, UK: Wiley-Blackwell.

Furst, P. T. (1990). *Flesh of the gods: The ritual use of hallucinogens*. Long Grove, IL: Waveland Press.

Gagan, J. M. (1998). *Journeying: Where psychology and shamanism meet*. Santa Fe, NM: Rio Chama Publications.

Gallagher, S. (2005). *How the body shapes the mind*. New York, NY: Oxford University Press.

Gallo, F. P. (ed., 2002). *Energy in psychotherapy*. New York, NY: W. W. Norton.

—— (2004). *Energy psychology: Explorations at the interface of energy, cognition, behavior, and health* (2nd ed.). London, UK: CRC/Taylor & Francis.

Gallop, J. (1988). *Thinking through the body*. New York, NY: Columbia University Press.

Gandhi, M. K. (2002). *The essential Gandhi: An anthology of his writings on his life, work, and ideas* (2nd ed.). New York, NY: Vintage.

Gendlin, E. T. (1982). *Focusing* (2nd ed.). New York NY: Bantam Books.

—— (1991). *Thinking beyond patterns: Body, language, and situations*. Berne, Switzerland: Peter Lang Publishing.

—— (1997). *Experiencing and the creation of meaning: A philosophical and psychological approach to the subjective*. Evanston, IL: Northwestern University Press.

—— (1998). *Focusing-oriented psychotherapy: A manual of the experiential method.* New York, NY: Guilford Press.

Gerber, R. (2000). *Vibrational medicine for the 21st century: The complete guide to energy healing and spiritual transformation.* New York, NY: HarperCollins.

—— (2001). *A practical guide to vibrational medicine: Energy healing and spiritual transformation.* New York, NY: Harper.

Gergen, M. M. & Davis, S. N. (eds, 1997). *Toward a new psychology of gender: A reader.* New York, NY: Routledge.

Gibran, K. (1969). *The prophet.* London, UK: Heinemann.

Gintis, B. (2007). *Engaging the movement of life: Exploring health and embodiment through osteopathy and continuum.* Berkeley, CA: North Atlantic Books.

Giorgi, A. (1970). *Psychology as a human science: A phenomenologically based approach.* New York, NY: Harper & Row.

—— (ed., 1985). *Phenomenology and psychological research.* Pittsburgh, PA: Duquesne University Press.

Glass, N. R. (1995). *Working emptiness: Toward a third reading of emptiness in Buddhism and postmodern thought.* Atlanta, GA: American Academy of Religion.

Glenn, L. & Müller-Schwefe, R. (eds, 1999). *The Radix reader.* Albuquerque, NM: Radix Institute.

Gleyse, J. (1997). *L'instrumentalisation du corps: Une archéologie de la rationalization instrumentale du corps, de l'âge classique a l'époque hypermoderne.* Paris, France: Harmattan.

Goethals, G. W. & Klos, D. S. (1976). *Experiencing youth: First-person accounts* (2nd ed.). Lanham, MD: University Press of America.

Gold, R. (2006). *Thai massage: A traditional medical technique* (2nd ed.). New York, NY: Mosby.

Goldstein, K. (1995). *The organism: A holistic approach to biology derived from pathological data in man* (new edition). New York: Zone Books.

Goodchild, W. (2001). *Eros and chaos: The sacred mysteries and dark shadows of love.* York Beach, ME: Nicolas-Hays.

Goodill, S. W. (2004). *An introduction to medical dance/movement therapy: Health-care in motion.* Philadelphia, PA: Jessica Kingsley.

Goodman, P. (1960). *Growing up absurd.* New York, NY: Vintage/Random House.

Goodson, A. (1991). *Therapy, nudity, and joy: The therapeutic use of nudity through the ages from ancient ritual to modern psychology* Los Angeles, CA: Elysium Growth Press.

Gopi, K. (1997). *Kundalini: The evolutionary energy in man.* Boston, MA: Shambala Press.

Graham, D. (2008). *A practical treatise on massage: Its history, mode of application, and effects.* Charleston, SC: Bibliolife.

Green, M. B. (1999). *Otto Gross, Freudian psychoanalyst, 1877–1920: Literature and ideas.* Lewiston, NY: Edwin Mellen Press.

Greene, B. (2004). *The fabric of the cosmos: Space, time, and the texture of reality.* New York, NY: Knopf.

Greenfield, S. A. (2001). *The private life of the brain: Emotions, consciousness, and the secret of the self.* New York, NY: Wiley.

Griffin, S. (1978). *Woman and nature: The roaring inside her.* New York, NY: Harper & Row.

—— (1996). *The eros of everyday life: Essays on ecology, gender and society.* New York, NY: Anchor Books.

Griffith, J. & Griffith, M. (1994). *The body speaks: Therapeutic dialogues for mind-body problems*. New York, NY: Basic Books.

Grigg, R. (2008). *Lacan, language, and philosophy*. Albany, NY: State University of New York.

Grisko, M. (ed., 1999). *Freikörperkultur und Lebenswelt*. Kassel, Germany: Kassel University Press.

Groddeck, G. (1961). *The book of the it*. New York, NY: Vintage Books.

—— (1988). *The meaning of illness: Selected psychoanalytic writings including correspondence with Sigmund Freud*. London, UK: Karnac Books.

Grof, S. (1988). *The adventure of self-discover: Dimensions of consciousness and new perspectives in psychotherapy*. Albany, NY: State University of New York Press.

—— (1998). *The cosmic game: Explorations on the frontiers of human consciousness*. Albany, NY: State University of New York Press.

—— (2000). *Psychology of the future: Lessons from modern consciousness research*. Albany, NY: State University of New York Press.

—— (ed., 2007). *Ancient wisdom and modern science*. Albany, NY: State University of New York Press.

—— & Bennet, H. Z. (1993). *The holotropic mind: The three levels of human consciousness and how they shape our lives*. New York, NY: HarperOne.

Gross, O. (2008). *Ausgewählte Texte 1901–1920*. Hamilton, NY: Mindpiece.

Grossman, D. (1996). *On killing: The psychological cost of learning to kill in war and society*. New York, NY: Back Bay Books.

Gruen, A. (2007). *The insanity of normality: Toward understanding human destructiveness*. Berkeley, CA: Human Development Books.

Guthrie, R. V. (1998). *Even the rat was white: A historical view of psychology* nd ed.). Boston, MA: Allyn & Bacon.

Habermas, J. (1972). *Knowledge and human interests* (trans. J. J. Shapiro). Boston, MA: Beacon Press.

Hackney, P. (2000). *Making connections: Total body integration through Bartenieff fundamentals*. New York, NY: Routledge.

Haldeman, S. (2004). *Principles and practices of chiropractic* (3rd ed.). New York, NY: McGraw-Hill Medical.

Hall, K. (2007). *The stuff of dreams: Fantasy, anxiety and psychoanalysis*. London, UK: Karnac Books.

Halprin, D. (2008). *The expressive body in life, art, and therapy: Working with movement, metaphor and meaning*. Philadelphia, PA: Jessica Kingsley.

Hameroff, S. R., Kaszniak, A. W. & Chalmers, D. J. (1999). *Toward a science of consciousness: The third Tucson discussion and debates*. Cambridge, MA: M.I.T. Press.

——, —— & Scott, A. C. (1998). *Toward a science of consciousness: The second Tucson discussion and debates*. Cambridge, MA: M.I.T. Press.

Hanh, T. N. (1999). *The miracle of mindfulness*. Boston, MA: Beacon Press.

—— (2008). *Breather, you are alive!: The sutra on the full awareness of breathing*. Berkeley, CA: Parallax Press.

Hanna, T. (1993). *The body of life: Creating new pathways for sensory awareness and fluid movement*. New York, NY: Knopf.

—— (2004). *Somatics: Reawakening the mind's control of movement, flexibility, and health*. Cambridge, MA: Da Capo Press.

Harding, S. (1998). *Is science multicultural? Postcolonialisms, feminisms, and epistemologies*. Bloomington, IN: Indiana University Press.

Harner, M. J. (ed., 1973). *Hallucinogens and shamanism*. London, UK: Oxford University Press.

—— (1984). *The Jivaro: People of the sacred waterfalls*. Berkeley, CA: University of California Press.

—— (1990). *The way of the shaman* (10ᵗʰ ed.). New York, NY: HarperOne.

Harrington, A. (2009). *The cure within: A history of mind-body medicine*. New York, NY: W. W. Norton.

Hart, T., Nelson, P. L. & Puhakka, K. (eds, 2000). *Transpersonal knowing: Exploring the horizon of consciousness*. Albany, NY: State University of New York Press.

Hartley, L. (1995). *Wisdom of the body moving: An introduction to body-mind centering*. Berkeley, CA: North Atlantic Books.

—— (2004). *Somatic psychology: Body, mind and meaning*. London, UK: Whurr Publishers.

Hayward, J. H. (1987). *Shifting worlds, changing minds: Where the sciences and Buddhism meet*. Boston, MA: Shambhala.

Hegel, G. W. F. (1977). *Phenomenology of spirit* (trans. A. V. Miller). Oxford, UK: Clarendon Press.

Heidegger, M. (1962). *Being and time* (trans., J. Macquarrie & E. Robinson). New York, NY: Harper and Row.

—— (1972). *On time and being* (trans., J. Stambaugh). New York, NY: Harper and Row.

—— (1982). *On the way to language*. New York, NY: HarperOne.

—— (2001). *Poetry, language, thought*. New York, NY: Harper Perennial.

—— (2008). *Basic writings*. New York, NY: Harper Perennial.

Heller, M. (ed., 2001). *The Flesh of the Soul: The body we work with: Selected papers of the 7ᵗʰ congress of the European Association of Body Psychotherapy*. Berne, Switzerland: Peter Lang Publishing.

Helminski, K. (2000). *The knowing heart: A Sufi path of transformation*. Boston, MA: Shambhala.

Henry, V. E. & Lifton, R. J. (2004). *Death work: Police, trauma, and the psychology of survival*. Oxford, UK: Oxford University Press.

Hergenhahn, B. R. (2008). *An introduction to the history of psychology* (6ᵗʰ ed.). Florence, KY: Wadsworth.

Hill, M. O. & Kandemwa, M. A. (2007). *Twin from another tribe: The story of two shamanic healers in Africa and North America*. Wheaton, IL: Quest Books.

Hillman, J. (1992). *The thought of the heart and the soul of the world*. Woodstock, CT: Spring Publications.

Holland, A. (2000). *Voices of Qi: An introductory guide to traditional Chinese medicine*. Berkeley, CA: North Atlantic Books.

Holzkamp, K. (1972). *Kritische Psychologie: Vorbereitende Arbeiten*. Frankfurt am Main, Germany: Fischer Verlag.

—— (1992). On doing psychology critically. *Theory and Psychology, 2*(2): 193–204.

Horkheimer, M. (1990). *Critical theory: Selected essays*. New York, NY: Seabury Press.

—— & Adorno, T. W. (2002). *Dialectic of enlightenment*. Palo Alto, CA: Stanford University Press.

Howitt, D. & Owusu-Bempah, J. (1994). *The racism of psychology: Time for change*. New York, NY: Harvester Wheatsheaf.

Huggins, M. K., Haritos-Fatouros, M. & Zimbardo, P. G. (2002). *Violence workers: Police torturers and murderers reconstruct Brazilian atrocities*. Berkeley, CA: University of California Press.

Husserl, E. (1960). *Cartesian meditations* (trans. D. Cairns). The Hague, Netherlands: Martinus Nijjhoff.

—— (1964). *The phenomenology of internal time-consciousness* (trans. J. S. Churchill). Bloomington, IN: Indiana University Press.

—— (1969). *Ideas: General introduction to pure phenomenology* (trans. W. R. B. Gibson). London, UK: Allen & Unwin.

—— (1970). *Logical investigations* (trans. J. N. Findlay). London, UK: Routledge & Kegan Paul, 1970.

—— (1974). *The crisis of European sciences and transcendental phenomenology* (trans. D. Carr). Evanston, IL: Northwestern University Press, 1974).

Hyde, L. (1983). *The gift: Imagination and the erotic life of property*. New York, NY: Vintage Books.

Ingerman, S. (2006). *Soul retrieval: Mending the fragmented self*. New York, NY: HarperOne.

Irigaray, L. (1985a). *Speculum of the other woman*. Ithaca, NY: Cornell University Press.

—— (1985b). *This sex which is not one*. Ithaca, NY: Cornell University Press.

Jacquette, D. (ed., 2004). *The Cambridge companion to Brentano*. London, UK: Cambridge University Press.

Jacoby, R. (1975). *Social amnesia: A critique of contemporary psychology from Adler to Laing*. Boston, MA: Beacon Press.

—— (1983). *The repression of psychoanalysis: Otto Fenichel and the political Freudians*. Chicago, IL: University of Chicago Press.

Jahnke, R. (2002). *The healing promise of qi: Creating extraordinary wellness through qigong and tai chi*. New York, NY: McGraw-Hill.

Jakobsen, M. D. (1999). *Shamanism: Traditional and contemporary approaches to the mastery of spirits and healing*. Oxford, UK: Berghahn Books.

Jameson, F. (1998). *The cultures of globalization*. Chapel Hill, NC: Duke University Press.

Jansz, J. & Drunen, P. V. (eds, 2003). *A social history of psychology*. Oxford, UK: Wiley-Blackwell.

Jarrett, J. L. (ed., 1997). *Jung's seminar of Nietzsche's Zarathustra*. Princeton, NJ: Princeton University Press.

Johanson, G. & Kurtz, R. (1991). *Grace unfolding: Psychotherapy in the spirit of the Tao-te Ching*. Bridgeport, PA: Bell Tower.

Johari, H. (2000). *Chakras: Energy centers of transformation*. Rochester, VT: Destiny Books.

Johnson, D. H. (1993). *Body: Recovering our sensual wisdom* (2nd ed.). Berkeley, CA: North Atlantic Books.

—— (1994). *Body, spirit, and democracy*. Berkeley, CA: North Atlantic Books.

—— (ed., 1995). *Bone, breath, and gesture: Practices of embodiment*. Berkeley, CA: North Atlantic Books.

—— (ed., 1997). *Groundworks: Narratives of embodiment*. Berkeley, CA: North Atlantic Books.

—— (2006). *Everyday hopes, utopian dreams: Reflections on American ideals*. Berkeley, CA: North Atlantic Books.

—— & Grand, I. (eds, 1998). *The body in psychotherapy: Inquiries in somatic psychology*. Berkeley, CA: North Atlantic Books.

Johnson, M. (1987). *The body in the mind: The bodily basis of meaning, imagination, and reason*. Chicago, IL: University of Chicago Press.

Johnson, R. (2001). *Elemental movement: A somatic approach to movement education*. Boca Raton, FL: BrownWalker Press.

Johnston, J. R. (ed., 2001). *The American body in context: An anthology*. Wilmington, DE: Scholarly Resources.

Joyce, J. (2006). *Dubliners*. Clayton, DE: Prestwick House.

Judith, A. (2004). *Eastern body, western mind: Psychology and the chakra system as a path to the self*. Berkeley, CA: Celestial Arts.

Jung, C. G. (1968a). Psychology and alchemy. *The collected works of Carl G. Jung*, Vol. 12. London, UK: Routledge & Kegan Paul.

—— (1968b). *Aion: Researches into the phenomenology of the self*. Princeton, NJ: Princeton University Press.

—— (1970). Symbols of transformation. *The collected works of Carl G. Jung*, Vol. 5. London, UK: Routledge & Kegan Paul.

—— (1971). Psychological types. *The collected works of Carl G. Jung*, Vol. 6. London, UK: Routledge & Kegan Paul.

—— (1981). *The archetypes and the collective unconscious* (2nd ed.; trans. G. Adler & R. F. C. Hull). Princeton, NJ: Princeton University Press.

—— (1988). *Nietzsche's Zarathustra: Notes of the seminar given in 1934–1939* (2 vols). Princeton, NJ: Princeton University Press.

—— (1999). *The psychology of kundalini yoga* (ed., S. Shamdasani). Princeton, NJ: Princeton University Press.

Kabat-Zinn, J. (1990). *Full catastrophe living: Using the wisdom of your body and mind to face stress, pain and illness*. New York, NY: Delta.

—— (2006). *Coming to our senses: Healing our selves and the world through mindfulness*. New York, NY: Hyperion Press.

—— (2007). *Arriving at your own door: 108 lessons in mindfulness*. New York, NY: Hyperion Press.

Kafatos, M. & Nadeau, R. (1990). *The conscious universe: Part and whole in modern physical theory*. New York, NY: Springer.

Kakar, S. (1982). *Shamans, mystics and doctors: A psychological inquiry into India and its healing traditions*. Chicago, IL: University of Chicago Press.

Kaltenbrunner, T. (2003). *Contact improvisation: Moving, dancing, interaction* (2nd ed.). Aachen, Germany: Meyer & Meyer Sport.

Kaminoff, L. (2007). *Yoga anatomy*. Champaign, IL: Human Kinetics Publishers.

Kandel, E. R., Schwartz, J. H. & Jessell, T. M. (2000). *Principles of neural science* (4th ed.). New York, NY: McGraw-Hill Medical.

Karagulla, S. & Gelder Kunz, D. v. G. (1989). *The chakras and the human energy fields*. Wheaton, IL: Theosophical Publishing House.

Kasulis, T. P., Ames, R. T. & Dissanayake, W. (eds, 1993). *Self as body in Asian theory and practice*. Albany, NY: State University of New York Press.

Kauffman, S. A. (1996). *At home in the universe: The search for the laws of self-organization and complexity*. London, UK: Oxford University Press.

—— (2002). *Investigations*. London, UK: Oxford University Press.

—— (2008) *Reinventing the sacred: A new view of science, reason, and religion*. New York, NY: Basic Books.

Keesling, B. (2006). *Sexual healing: The complete guide to overcoming common sexual problems* (3ʳᵈ ed.). Alameda, CA: Hunter House.

Kehoe, A. B. (2000). *Shamans and religion: An anthropological exploration in critical thinking.* Long Grove, IL: Waveland Press.

Keleman, S. (1975a). *Your body speaks its mind.* Berkeley, CA: Center Press.

—— (1975b). *Human ground.* Berkeley, CA: Center Press.

—— (1986). *Emotional anatomy: The structure of experience.* Berkeley, CA: Center Press.

—— (1987a). *Bonding: A somatic-emotional approach to transference.* Berkeley, CA: Center Press.

—— (1987b). *Embodying experience: Forming a personal life.* Berkeley, CA: Center Press.

—— (1999). *Myth and the body: A colloquy with Joseph Campbell.* Berkeley, CA: Center Press.

Keller, E. F. (1985). *Reflections on gender and science.* New Haven, CT: Yale University Press.

Kelley, C. R. (1974). *Education in feeling and purpose* (2ⁿᵈ ed.). Santa Monica, CA: Radix Institute.

—— (2004). *Life force: The creative process in man and nature.* Victoria, BC, Canada: Trafford Press.

Khalsa, D. F. & O'Keeffe, D. (2002). *The kundalini yoga experience: Bringing body, mind, and spirit together.* New York, NY: Fireside.

Khan, P. V. I. (2000). *Awakening: A Sufi experience.* New York, NY: Tarcher.

Klein, M. (1921–1963). *The writings of Melanie Klein* (4 vols). London, UK: Hogarth, 1975.

Kleinberg-Levin, D. M. (1985). *The body's recollection of being: Phenomenological psychology and the deconstruction of nihilism.* London, UK: Routledge & Kegan Paul.

—— (2005). *Gestures of ethical life: Reading Hölderlin's question of measure after Heidegger.* Palo Alto, CA: Stanford University Press.

—— (2009). *Before the voice of reason: Echoes of responsibility in Merleau-Ponty's ecology and Levinas' ethics.* Albany, NY: State University of New York Press.

Kleinman, A. (1988). *Rethinking psychiatry: From cultural category to personal experience.* New York, NY: Collier Macmillan.

Knaster, M. (1996). *Discovering the body's wisdom.* New York, NY: Bantam.

Koch, C. (2004). *The quest for consciousness: A neurobiological approach.* Greenwood, Village, CO: Roberts & Company.

Kohák, E. (1984). *The embers and the stars: A philosophical inquiry into the moral sense of nature.* Chicago, IL: University of Chicago Press.

Kolakowski, L. (1988). *Metaphysical horror.* New York, NY: Basil Blackwell.

—— (1989). *The presence of myth* (trans. A. Czerniawski). Chicago, IL: University of Chicago Press.

Kolbert, E. (2006). *Field notes from a catastrophe: Man, nature, and climate change.* London, UK: Bloomsbury Publishing.

Kosslyn, S. M. (1980). *Image and mind.* Cambridge, MA: Harvard University Press.

Kring, A. M., Davison, G. C., Neale, J. M. & Johnson, S. L. (2006). *Abnormal psychology* (10ᵗʰ ed.). New York, NY: Wiley.

Kripal, J. (2007). *Esalen: America and the religion of no religion.* Chicago, IL: University of Chicago Press.

—— & Shuck, G. W. (eds, 2005). *On the edge of the future: Esalen and the evolution of American culture.* Bloomington, IN: Indiana University Press.

Krippner, S. (1992). *Spiritual dimensions of healing: From native shamanism to contemporary health care.* New York, NY: Irvington Publishing.

Krishna, G. (1997. *Kundalini: The evolutionary energy in man.* Boston, MA: Shambhala.

Krishnamurti, J. (1975). *Freedom from the known.* San Francisco, CA: HarperOne.

—— (1996). *Total freedom: The essential Krishnamurti.* San Francisco, CA: HarperOne.

—— (2007). *As one is: To free the mind from all conditioning.* Prescott, AZ: Hohm Press.

Kristeva, J. (1975). *The system and the speaking subject.* Lisse, Netherlands: Peter de Ridder.

—— (1980). *Desire in language.* New York, NY: Columbia University Press.

—— (1982). *Powers of horror: An essay on abjection.* New York, NY: Columbia University Press.

—— (1984). *Revolution in poetic language.* New York, NY: Columbia University Press.

—— (1987). *In the beginning was love: Psychoanalysis and faith.* New York, NY: Columbia University Press.

Kubler-Ross, E. (1997). *On death and dying.* New York, NY: Scribner.

Kuefler, M. (ed., 2007). *The history of sexuality sourcebook.* Peterborough, Canada: Broadview Press.

Kugle, S. (2007). *Sufis and saints' bodies: Mysticism, corporeality, and sacred power in Islam.* Chapel Hill, NC: University of North Carolina Press.

Kuriyama, S. (2002). *The expressiveness of the body and the divergence of Greek and Chinese medicine.* New York, NY: Zone Books.

Kurtz, R. (2007). *Body-centered psychotherapy.* Mendocino, CA: LifeRhythm.

—— & Prestera, H. (1984). *Body reveals: How to read your own body.* New York, NY: HarperCollins.

Kushi, M. (2007). *Your body never lies: The complete book of oriental diagnosis.* Garden City Park, NY: Square One Publishers.

Kvale, S. (ed., 1992). *Psychology and postmodernism.* London, UK: Sage.

Lacan, J. (1972). *Écrits: A selection* (trans. A. Sheridan). London, UK: Tavistock.

—— (1977). *The four fundamental concepts of psychoanalysis* (trans. A. Sheridan). London: Hogarth.

Lad, V. (2001). *Textbook of āyurveda, volume one: Fundamental principles.* Albuquerque, NM: Āyurvedic Press.

—— (2007). *Textbook of āyurveda, volume two: A complete guide to clinical assessment.* Albuquerque, NM: Āyurvedic Press.

Laing, R. D. (1960). *The divided self: An existential study in sanity and madness.* London, UK: Penguin Books.

—— (1983). *The politics of experience.* New York, NY: Pantheon.

Lakoff, G. & Johnson, M. (1980). *Metaphors we live by.* Chicago, IL: University of Chicago Press.

—— & —— (1999). *Philosophy in the flesh: The embodied mind and its challenge to western thought.* New York, NY: Basic Books.

Lawson, R. B., Graham, J. E. & Baker, K. M. (2006). *A history of psychology.* Upper Saddle River, NJ: Prentice Hall.

Leahey, T. H. (2003). *A history of psychology: Main currents in psychological thought* th ed.). Upper Saddle River, NJ: Prentice Hall.

Leete, A. & Firnhaber, R. P. (eds, 2004). *Shamanism in the interdisciplinary context.* Boca Raton, FL: Brown Walker Press.

Lefebvre, H. (2008). *The critique of everyday life* (3 vols). London, UK: Verso.

Lethbridge, D. (1991). *Mind in the world: The Marxist psychology of self-actualization.* Minneapolis, MN: MEP Publishing.

Lévi-Strauss, C. (1963). *Structural anthropology.* New York, NY: Basic Books.

Levinas, E. (1969). *Totality and infinity: An essay on exteriority* (trans. A. Lingis). Pittsburgh, PA: Duquesne University Press.

—— (1990). *Time and the other* (new ed.; trans. R. A. Cohen). Pittsburgh, PA: Duquesne University Press.

—— (1998). *Otherwise than being: Or beyond essence.* Pittsburgh, PA: Duquesne University Press.

—— (2005). *Humanism of the other.* Champaign, IL: University of Illinois Press.

—— & Cohen, R. A. (1990). *Time and the other.* Pittsburgh, PA: Duquesne University Press.

—— & Smith, M. B. (1999). *Alterity and transcendence.* New York, NY: Columbia University Press.

Levine, P. A. (2008). *Healing trauma: A pioneering program for restoring the wisdom of your body.* Boulder, CO: Sounds True.

—— & Frederick, A. (1997). *Waking the tiger — Healing trauma: The innate capacity to transform overwhelming experiences.* Berkeley, CA: North Atlantic Books.

Lewin, B. D. (1933). The body as phallus. *Psychoanalytic Quarterly, 2*: 24–47.

Lewis, P. (1994). *Theoretical approaches in dance-movement therapy.* Dubuque, IA: Kendall Hunt Publishing.

—— (ed., 2002). *Integrative holistic health, healing, and transformation: A guide for practitioners, consultants, and administrators.* Springfield, IL: Charles C. Thomas.

Liang, S-Y. & Wu, W-C. (1996). *Qigong empowerment: A guide to medical, Taoist, Buddhist, Wushu energy cultivation.* East Providence, RI: Way of the Dragon Press.

Lifton, R. J. (1976). *Boundaries: Psychological man in revolution.* New York, NY: Vintage.

—— (1986). *The Nazi doctors: Medical killing and the psychology of genocide.* New York, NY: Basic Books.

—— (1993). *The protean self: Human resilience in an age of fragmentation.* New York, NY: Basic Books.

Limentani, A. (1989). *Between Freud and Klein: The psychoanalytic quest for knowledge and truth.* London, UK: Free Association Books.

Lingis, A. (1983). *Excesses: Eros and culture.* Albany, NY: State University of New York Press.

—— (1985). *Libido: French existential theories.* Bloomington, IN: Indiana University Press.

—— (1989). *Deathbound subjectivity.* Bloomington, IN: Indiana University Press.

—— (1994). *Foreign bodies.* New York, NY: Routledge.

—— (1996). *Sensation: Intelligibility in sensibility.* Atlantic Highlands, NJ: Humanity Books.

—— (2005). *Body transformations: Evolution and atavism in culture.* New York, NY: Routledge.

Littlewood, W. C. & Roche, M. A. (2004). *Waking up: The work of Charlotte Selver.* Bloomington, IN: Author House.

Llinas, R. (2001). *I of the vortex: From neurons to self.* Cambridge, MA: M.I.T. Press.

Lock, M. & Farquhar, J. (2007). *Beyond the body proper: Reading the anthropology of material life.* Durham, NC: Duke University Press.

Louv, R. (2008). *Last child in the woods: Saving our children from nature-deficit disorder.* Chapel Hill, NC: Algonquin Books.

Lowen, A. (1965). *Love and orgasm.* New York, NY: Macmillan.

—— (1976). *The language of the body.* [Originally published in 1958 as *The physical dynamics of character structure.*] New York, NY: Collier.

—— (1990). *The spirituality of the body: Bioenergetics for grace and harmony.* New York, NY: Macmillan.

—— (1994). *Bioenergetics: The revolutionary therapy that uses the language of the body to heal the problems of the mind.* London, UK: Penguin Arkana.

—— (2003). *Fear of life.* Alachua, FL: Bioenergetics Press.

—— (2005a). *The voice of the body.* Alachua, FL: Bioenergetics Press.

—— (2005b). *The betrayal of the body: The psychology of fear and terror* (3rd ed.). Alachua, FL: Bioenergetics Press.

Lovelock, J. (2007). *The revenge of Gaia: Earth's climate crisis and the fate of humanity* (new ed.). New York: Basic Books.

Lu, H. C. (2005). *Traditional Chinese medicine: An authoritative and comprehensive guide.* Laguna Beach, CA: Basic Health Publications.

Macherey, P. (1998). *In a materialist way: Selected essays.* London, UK: Verso.

—— (2006). *A theory of literary production.* New York, NY: Routledge.

Macnaughton, I. (ed., 2004). *Body, breath, and consciousness: A somatics anthology.* Berkeley, CA: North Atlantic Books.

Maddox, J. L. (2003). *Shamans and shamanism.* New York, NY: Dover Publications.

—— & Keller, A. G. (2003). *Medicine man: A sociological study of the character and evolution of shamanism.* Whitefish, MT: Kessinger Publishing.

Main, T. (1989). *The ailment and other psychoanalytic essays.* London, UK: Free Association Books.

Mann, D. (ed., 1999). *Erotic transference and countertransference: Clinical practice in psychotherapy.* New York, NY: Routledge.

Marcuse, H. (1971). *An essay on liberation.* Boston, MA: Beacon Press.

—— (1987). *Eros and civilization: A philosophical inquiry into Freud* (2nd ed.). New York, NY: Routledge.

—— (2006). *One-dimensional man: Studies in the ideology of advanced industrial society.* New York, NY: Routledge.

Margulis, L. & Sagan, D. (1986). *Microcosmos: Four billion years of microbial evolution.* Berkeley, CA: University of California Press.

Marlock, G. & Weiss, H. (2006). *Handbuch der Körperpsychotherapie.* Stuttgart, Germany: Schattauer.

Martin, A. & Landrell, J. (2005). *Energy psychology, energy medicine.* Penryn, CA: Personal Transformation Press.

Martin, J. (2000). *Wilhelm Reich and the cold war.* Mendocino, CA: Flatland Books.

Martín-Baró, I. (1994). *Writings for a liberation psychology*. Cambridge, MA: Harvard University Press.

Maslow, A. H. (1970). *Religions, values, and peak experiences*. New York, NY: Viking Press.

Maybury-Lewis, D. (1992). *Millennium: Tribal wisdom and the modern world*. Darby, PA: Diane Publishing Company.

—— (2001). *Indigenous peoples, ethnic groups, and the state* (2nd ed.). Boston, MA: Allyn and Bacon.

McCarty, W. A. (2008). *La conciencia del bebe antes de nacer: Un comienzo milagroso*. Col. Santa Cruz Atoyac, México: Editorial Pax México.

McLaren, P. & Lankshear, C. (eds, 1994). *Politics of liberation: Paths from Freire*. New York, NY: Routledge.

McLuhan, H. M. (1964). *Understanding media: The extensions of man*. New York, NY: McGraw Hill.

McGowan, D. (1997). *Alexander technique: Original writings of F. M. Alexander*. Burdett, NY: Larson Publications.

Mchose, C. & Frank, K. (2006). *How life moves: Explorations in meaning and body awareness*. Berkeley, CA: North Atlantic Books.

McKeachie, W. J. (1976). Psychology in America's bicentennial year. *American Psychologist, 31*: 819–833.

McLaren, P. (1994). *Critical pedagogy and predatory culture: Oppositional politics in a postmodern era*. London, UK: Routledge.

—— & Kincheloe, J. L. (eds, 2007). *Critical pedagogy: Where are we now?* Berne, Switzerland: Peter Lang Publishing.

McNeely, D. A. (1987). *Touching: Body therapy and depth psychology*. Toronto, CA: Inner City Books.

McNely, J. K. (1981). *Holy wind in Navajo philosophy*. Tucson, AZ: University of Arizona Press.

Meckel, J. & Moore, R. L. (eds, 1992). *Self and liberation: The Jung/Buddhism dialogue*. New York, NY: Paulist Press.

Meekums, B. (2002). *Dance movement therapy: A creative psychotherapeutic approach*. London, UK: Sage.

Memmi, A. (1991). *The colonizer and the colonized*. Boston, MA: Beacon Press.

—— (2000). *Racism*. Minneapolis, MN: University of Minnesota Press.

—— (2006). *Decolonization and the decolonized*. Minneapolis, MN: University of Minnesota Press.

Merchant, C. (1980). *The death of nature: Women, ecology, and the scientific revolution*. San Francisco, CA: Harper & Row.

Merleau-Ponty, M. (1962). *Phenomenology of perception* (trans., C. Smith). London, UK: Routledge and Kegan Paul.

—— (1963). *The structure of behavior* (trans., A. Fisher). Boston, MA: Beacon Press.

—— (1964). *Sense and non-sense* (trans., H. Dreyfus & P. A. Dreyfus). Evanston, IL: Northwestern University Press.

—— (1968). *The visible and the invisible* (trans., A. Lingis). Evanston, IL: Northwestern University Press.

—— (1973a). *Consciousness and the acquisition of language* (trans., H. J. Silverman). Evanston, IL: Northwestern University Press.

—— (1973b). *The prose of the world* (trans., J. O'Neil). Evanston, IL: Northwestern University Press.

Metzner, R. (1999). *Green psychology: Transforming our relationship to the earth.* Rochester, VT: Park Street Press.

Miles, P. (2008). *Reiki: A comprehensive guide.* New York, NY: Tarcher.

Mills, C. W. (1997). *The racial contract.* Ithaca, NY: Cornell University Press.

Mills, J. A. (2000). *Control: A history of behavioral psychology.* New York, NY: New York University Press.

Mindell, A. P. (1982). *Dreambody, the body's role in revealing the self.* Santa Monica, CA: Sigo Press.

—— (1985a). *River's way: The process science of the dreambody.* London, UK: Routledge & Kegan Paul.

—— (1985b). *Working with the dreaming body.* London, UK: Routledge & Kegan Paul.

—— (1987). *The dreambody in relationships.* London: Routledge & Kegan Paul.

—— (1988). *City shadows: Psychological interventions in psychiatry.* London, UK: Routledge.

—— (1991a). *The personal and global dreambody.* San Francisco, CA: New Dimensions Foundation.

—— (1991b). *Your body speaks its dream.* San Francisco, CA: New Dimensions Foundation.

—— (1992). *Riding the horse backwards: Process work in theory and practice.* New York, NY: Arkana.

—— (1993). *The shaman's body: A new shamanism for transforming health, relationships, and community.* San Francisco, CA: Harper.

—— (2000). *Quantum mind: The edge between physics and psychology.* Portland, OR: Lao Tse Press.

—— (2007). *Earth-based psychology: Path awareness from the teachings of Don Juan, Richard Feynman, and Lao Tse.* Portland, OR: Lao Tse Press.

Minow, M. (1998). *Between vengeance and forgiveness: Facing history after genocide and mass violence.* Boston, MA: Beacon Press.

Minuchin, S., Rosman, B. L. & Baker, L. (2004). *Psychosomatic families: Anorexia nervosa in context.* Cambridge, MA: Harvard University Press.

Mio, J., Barker-Hackett, L. & Tumambing, J. (2008). *Multicultural psychology: Understanding our diverse communities* (2nd ed.). New York, NY: McGraw-Hill.

Mollon, P. (2005). *EMDR and the energy therapies: Psychoanalytic perspectives.* London, UK: Karnac Books.

—— (2008). *Psychoanalytic energy psychotherapy: Inspired by thought field therapy, EFT, TAT, and Seemorg Matrix.* London, UK: Karnac Books.

Montagu, A. (1971). *Touching: The human significance of skin* (3rd ed.). New York, NY: Harper & Row.

Mookerjee, A. (1981). *Kundalini: The arousal of the inner energy* (2nd ed.). Rochester, VT: Destiny Books.

Morin, E. (2008). *On complexity.* Cresskill, NJ: Hampton Press.

Morris, D. R. (1991). *The culture of pain.* Berkeley, CA: University of California Press.

Moss, D. (ed., 1999). *Humanistic and transpersonal psychology: A historical and biographical sourcebook.* Westport, CT: Greenwood Press.

Mruk, C. J. & Hartzell, J. (2007). *Zen and psychotherapy: Integrating traditional and nontraditional approaches.* New York, NY: Springer.

Mullen, B. & Laing, R. D. (1996). *Mad to be normal: Conversations with R. D. Laing.* London, UK: Free Association Books.

Muller, J. P. & Tillman J. G. (eds, 2007). *The embodied subject: Minding the body in psychoanalysis.* Lanham, MD: Jason Aronson.

Murphy, M. (1992). *The future of the body: Explorations into the further evolution of human nature.* New York, NY: Tarcher/Putnam.

Murphy, N. (2006). *Bodies and souls, or spirited bodies?* London, UK: Cambridge University Press.

Myss, C. (1996). *Anatomy of the spirit: The seven stages of power and healing.* New York, NY: Three Rivers Press.

Naranjo, C. (2006a). *The way of silence and the talking cure: On meditation and psychotherapy.* Nevada City, CA: Blue Dolphin Publishing.

—— (2006b). *The one quest: A map of the ways of transformation.* Nevada City, CA Gateways Books.

Narby, J. & Huxley, F. (eds, 2001). *Shamans through time: 500 years on the path to knowledge.* New York, NY: Tarcher/Putnam.

Natoli, J. & Hutcheon, L. (eds, 1993). *A postmodern reader.* Albany, NY: State University of New York.

Nauriyal, D. K., Drummond, M. & Lal, Y. B. (eds, 2006). *Buddhist thought and applied psychological research: Transcending the boundaries.* London, UK: Routledge.

Nettle, D. & Romaine, S. (2000). *Vanishing voices: The extinction of the world's languages.* London, UK: Oxford University Press.

Nettleton, S. (2006). *The sociology of health and illness* (2nd ed.). Cambridge, UK: Polity.

Newberg, A., D'Aquili, E. & Rause, V. (2002). *Why god won't go away: Brain science and the biology of belief.* New York, NY: Ballantine.

Ninivaggi, F. J. (2008). *Āyurveda: A comprehensive guide to traditional Indian medicine for the west.* Westport, CT: Praeger.

Niranjanananda, Sw. & Niranjanananda, Sw. S. (2002). *Prana pranayama prana vidya.* Coimbatore, India: Yoga Publishing Trust.

Noë, A. (2006). *Action in perception.* Cambridge, MA: MIT Press.

—— (2009). *Out of our heads: Why you are not your brain, and other lessons from the biology of consciousness.* New York, NY: Hill and Wang.

Noel, D. C. (1999). *The soul of shamanism: Western fantasies, imaginal realities.* London, UK: Continuum.

Northoff, G. (2004). *Philosophy of the brain.* Amsterdam, Netherlands: John Benjamins Publishing.

Novalis. (1997). *Novalis: Philosophical writings.* New York, NY: State University of New York Press.

Ohashi. (1991). *Reading the body: Ohashi's book of oriental diagnosis.* New York, NY: Penguin Putnam.

Ogden, P., Minton, K. & Pain, C. (2006). *Trauma and the body: A sensorimotor approach to psychotherapy.* New York, NY: W. W. Norton.

Oliver, K. & Edwin, S. (eds, 2002). *Between the psyche and the social: Psychoanalytic social theory.* Lanham, MD: Rowman & Littlefield.

Olsen, A. & McHose, C. (2004). *Bodystories: A guide to experiential anatomy.* Hanover, NH: UPNE Publishing.

Ornstein, R. & Sobel, D. (1988). *The healing brain.* New York, NY: Simon & Schuster.

Oschman, J. L. (2000). *Energy medicine: The scientific basis.* New York, NY: Churchill Livingstone.

Osho (2004a). *Freedom: The courage to be yourself.* London, UK: St. Martin's Griffin.

—— (2004b). *Meditation: The first and last freedom.* London, UK: St. Martin's Griffin.

—— (2005). *Body mind balancing: Using your mind to heal your body.* London, UK: St. Martin's Griffin.

—— (2009). *Joy: The happiness that comes from within.* London, UK: St. Martin's Griffin.

Otto, R. (2004). *The idea of the holy: An inquiry into the non-rational factor in the idea of the divine.* Whitefish, MT: Kessinger Publishing.

Painter, J. W. (1984). *Deep bodywork and personal development: Harmonizing our bodies, emotions and thoughts.* Mill Valley, CA: Center for Release and Integration.

Pallant, C. (2006). *Contact improvisation: An introduction to a vitalizing dance form.* Jefferson, NC: McFarland and Company.

Pallaro, P. (ed., 1999). *Authentic movement: Essays by Mary Starks Whitehouse, Janet Adler and Joan Chodorow.* Philadelphia, PA: Jessica Kingsley Publishers.

—— (2007). *Authentic movement: Moving the body, moving the self, being moved.* Philadelphia, PA: Jessica Kingsley Publishers.

Palmer, D. D. (2006). *Chiropractic: A science, an art and the philosophy thereof.* Whitefish, MT: Kessinger Publishing.

Paranjpe, A. C. (1984). *Theoretical psychology: The meeting of east and west.* New York, NY: Springer.

Paris, G. (2007). *Wisdom of the psyche: Depth psychology after neuroscience.* New York, NY: Routledge.

Park, J. Y. (2006). *Buddhisms and deconstructions.* Lanham, MD: Rowman & Littlefield.

Parker, I. & Spears, R. (eds, 1996). *Psychology and society: Radical theory and practice.* London, UK: Pluto.

Payne, H. (2006). *Dance movement therapy: Theory, research and practice* (2nd ed.). London, UK: Routledge.

Pearson, B. A. (2007). *Ancient Gnosticism: Traditions and literature.* Minneapolis, MN: Fortress Press.

Peerbolte, M. L. (1975). *Psychic energy.* Wassenaar, Netherlands: Servire.

Penrose, R. (2007). *The road to reality: A complete guide to the laws of the universe.* New York, NY: Vintage.

Perce, M., Forsythe, A. M. & Ball, C. (1992). *The dance technique of Lester Horton.* Princeton, NJ: Princeton Book Company.

Perls, F. (1969). *Ego, hunger and aggression: The beginning of gestalt therapy.* New York, NY: Vintage.

——, Hefferline, R. F. & Goodman, P. (1951). *Gestalt therapy: Excitement and growth in the human personality.* New York, NY: The Julian Press.

Perkins, J. (2005). *Confessions of an economic hitman.* New York, NY: Plume.

Pert, C. (1997). *Molecules of emotion.* New York, NY: Scribner.

Pesso, A. (1969). *Movement in psychotherapy: Psychomotor techniques and training.* New York, NY: New York University Press.

——, (1990). *Moving psychotherapy: Theory and application of the Pesso system of psychomotor therapy*. Brookline, MA: Brookline Books.

Petitot, J., Varela, F., Pachoud, B. & Roy, J-M. (2000). *Naturalizing phenomenology: Issues in contemporary phenomenology and cognitive science*. Palo Alto, CA: Stanford University Press.

Piano, F. D., Olson, R. P., Mukherjee, A., Kamilar, S. M., Hagen, L. & Hartsman, E. (2002). *Religious theories of personality and psychotherapy: East meets west*. New York, NY: Routledge.

Pierrakos, E. (1993). *Fear no evil: The pathwork method of transforming the lower self*. Charlottesville, VA: Pathwork Press.

Pierrakos, J. C. (2001). *Eros, love and sexuality: The forces that unify man and woman*. Mendocino, CA: LifeRhythm.

—— (2005). *Core energetics: Developing the capacity to love and heal* (2nd ed.). Mendocino, CA: Core Evolution Publishing.

Pinchbeck, D. (2003). *Breaking open the head: A psychedelic journey into the heart of contemporary shamanism*. New York, NY: Broadway.

Piontelli, A. (1986). *Backwards in time: A study in infant observation by the method of Esther Bick*. London, UK: Karnac.

Plotkin, B. (2007). *Nature and the human soul: Cultivating wholeness and community in a fragmented world*. Novato, CA: New World Library.

—— & Berry, T. (2003). *Soulcraft: Crossing into the mysteries of nature and psyche*. Novato, CA: New World Library.

Plotnik, R. (2005). *Introduction to psychology*. Belmont, CA: Thomson-Wadsworth.

Pole, S. (2006). *Āyurvedic medicine: The principles of traditional practice*. Edinburgh, UK: Churchill Livingstone.

Polhemous, T. (ed., 1978). *The body reader: Social aspects of the human body*. New York, NY: Pantheon Books.

Porter, R. (2003). *Madness: A brief history*. Oxford, UK: Oxford University Press.

Powers, W. K. (1986). *Sacred language: The nature of supernatural discourse in Lakota*. Norman, OK: University of Oklahoma Press.

Price, J. & Shildrick, M. (eds, 1999). *Feminist theory and the body: A reader*. New York, NY: Routledge.

Price, N. (2001). *The archaeology of shamanism*. New York, NY: Routledge.

Pratt, C. (2007). *An encyclopedia of shamanism* (2 vols). New York, NY: Rosen Publishing.

Prilleltensky, I. (1992). Humanistic psychology, human welfare and the social order. *The Journal of Mind and Behaviour, 13*(4): 315–327.

—— & Nelson, G. (2002). *Doing psychology critically: Making a difference in diverse settings*. New York, NY: Palgrave Macmillan.

Purves, D. (2007). *Neuroscience* (4th ed.). Sunderland, MA: Sinauer Associates.

Qualls-Corbett, N. (1988). *The sacred prostitute: Eternal aspect of the feminine*. Toronto, Canada: Inner City Books.

Rabaté, J-M. (ed., 2003). *The Cambridge companion to Lacan*. London, UK: Cambridge University Press.

Rachlin, H. (1991). *Introduction to modern behaviorism* (3rd ed.). New York, NY: Freeman.

Radin, D. (1997). *The conscious universe: The scientific truth of psychic phenomena*. New York, NY: HarperOne.

—— (2006). *Entangled minds: Extrasensory experiences in a quantum reality.* New York, NY: Paraview Pocket Books.

Raknes, O. (2004). *Wilhelm Reich and orgonomy.* Princeton, NJ: American College of Orgonomy Press.

Rama, S., Ajaya, S. & Ballentine, R. (1976). *Yoga and psychotherapy: The evolution of consciousness.* Honesdale, PA: Himalayan Institute Press.

Ramachandran, V. S. (2003). *The emerging mind: The Reith lectures.* London, UK: Profile Books.

—— (2005). *A brief tour of human consciousness.* Los Angeles, CA: Pi Books.

Ramacharaka, Y. (2006). *The secret of prana.* Whitefish, MT: Kessinger Publishing.

Ramos, D. G. (2004). *The psyche of the body: A Jungian approach to psychosomatics.* New York, NY: Brunner-Routledge.

Ramsdale, D. A. & Dorfman, E. J. (1985). *Sexual energy ecstasy.* Playa Del Rey, CA: Peak Skill Publishing.

Rank, O. (1994). *The trauma of birth* (2nd ed.). New York, NY: Dover Publications.

—— (1996). *A psychology of difference: The American lectures* (trans., R. Kramer). Princeton, NJ: Princeton University Press.

Rao, R. S. K. (1982). *Śrī Cakra: Its yantra, mantra, and tantra.* Bangalore, India: Kalpatharu Resarch Academy.

Ray, R. A. (2002a). *Indestructible truth: The living spirituality of Tibetan Buddhism.* Boston, MA: Shambhala.

—— (2002b). *Secret of the vajra world: The tantric Buddhism of Tibet.* Boston, MA: Shambhala.

—— (2008). *Touching enlightenment: Finding realization in the body.* Louisville, CO: Sounds True.

Reich, A. (1953). Narcissistic object choice in women. *Journal of the American Psychoanalytic Association, 1*: 22–44.

Reich, W. (1949). Preface to the third edition. In: *Character analysis* (trans. T. P. Wolfe). New York, NY: Noonday Press, pp. ix–xi.

—— (1961). *Wilhelm Reich selected writings: An introduction to orgonomy* (ed., M. Higgins). New York, NY: Farrar, Straus and Giroux.

—— (1967). *Reich speaks of Freud.* New York, NY: Farrar, Straus and Giroux.

—— (1971a). *The sexual revolution: Toward a self-governing character structure.* New York, NY: Farrar, Straus and Giroux.

—— (1971b). *The invasion of compulsory sex-morality.* New York, NY: Farrar, Straus and Giroux.

—— (1974). *Listen, little man!* New York, NY: Farrar, Straus and Giroux.

—— (1979). *The bion experiments.* New York, NY: Farrar, Straus and Giroux.

—— (1980a). *Character analysis* (3rd enlarged ed.). New York, NY: Farrar, Straus and Giroux.

—— (1980b). *Genitality: In the theory and therapy of neurosis.* New York, NY: Farrar, Straus and Giroux.

—— (1980c). *The mass psychology of fascism.* New York, NY: Farrar, Straus and Giroux.

—— (1986). *The function of the orgasm: The discovery of the orgone.* New York, NY: Farrar, Straus and Giroux.

Reinharz, S. (1992). *Feminist methods in social research.* New York, NY: Oxford University Press.

Reiss, I. L. (2006). *An insider's view of sexual science since Kinsey*. Lanham, MD: Rowman & Littlefield.

Reiss, T. J. (1982). *The discourse of modernism*. Ithaca, NY: Cornell University Press.

—— (1988). *The uncertainty of analysis: Problems in truth, meaning, and culture*. Ithaca, NY: Cornell University Press.

—— (1997). *Knowledge, discovery and imagination in early modern Europe: The rise of aesthetic rationalism*. London, UK: Cambridge University Press.

—— (2002a). *Against autonomy: Global dialectics of cultural exchange*. Palo Alto, CA: Stanford University Press.

—— (2002b). *Mirages of the self: Patterns of personhood in ancient and early modern Europe*. Palo Alto, CA: Stanford University Press.

Rey, R. (1995). *The history of pain*. Cambridge, MA: Harvard University Press.

Reynolds, N. & McCormick, M. (2003). *No fixed points: Dance in the twentieth century*. New Haven, CT: Yale University Press.

Richards, G. (1997). *"Race," racism and psychology: Towards a reflexive history*. London, UK: Routledge.

Ricoeur, P. (1967). *Husserl: An analysis of his phenomenology* (trans., E. G. Ballard & L. E. Embree). Evanston, IL: Northwestern University Press.

—— (1970). *Freud and philosophy: An essay on interpretation*. New Haven, CT: Yale University Press.

—— (1974). *The conflict of interpretations: Essays in hermeneutics* (ed., D. Ihde). Evanston, IL: Northwestern University Press.

Rispoli, L. (1993). *Psicologia Funzionale del Sé*. Roma, Italy: Astrolabio.

Robinson, P. A. (1990). *The Freudian Left: Wilhelm Reich, Geza Roheim, Herbert Marcuse*. Ithaca, NY: Cornell University Press.

Rogat, S. (1997). *Healing thoughts, therapeutic shamanism: A bridge between metaphysics and psychotherapy*. Parker, CO: Paintbrush Press.

Rogers, C. (1989). *The Carl Rogers reader*. Boston, MA: Mariner Books.

Rolf, I. P. (1989). *Rolfing: Reestablishing the natural alignment and structural integration of the human body for vitality and well-being*. Rochester, VT: Healing Arts Press.

—— (1990). *Rolfing and physical reality*. Rochester, VT: Healing Arts Press.

Romanyshyn, R. D. (2002). *Ways of the heart: Essays toward an imaginal psychology*. Amherst, NY: Trivium Publications.

Rose, N. (1996). *Inventing our selves: Psychology, power, and personhood*. Cambridge, UK: Cambridge University Press.

Rosenau, P. M. (1992). *Post-modernism and the social sciences: Insights, inroads, and intrusions*. Princeton, NJ: Princeton University Press.

Rosenberg, J. L. (1973). *Total orgasm*. New York, NY: Random House/Bookworks.

——, Rand, M. L. & Asay, D. (1987). *Body, self, and soul: Sustaining integration*. Lake Worth, FL: Humanics Press.

Rosiello, F. W. (2000). *Deepening intimacy in psychotherapy: Using the erotic transference and countertransference*. Lanham, MD: Jason Aronson.

Roszak, T. (2001). *The voice of the earth: Explorations in ecopsychology* (2nd ed.). Glasgow, UK: Phanes Press.

——, Gomes, M. E. & Kanner, A. D. (eds, 1995). *Ecopsychology: Exploring the earth, healing the mind*. San Francisco, CA: Sierra Club Books.

Rothschild, B. (2000). *The psychophysiology of trauma and trauma treatment.* New York, NY: W. W. Norton.

—— (2003). *The body remembers casebook: Unifying methods and models in the treatment of trauma and PTSD.* New York, NY: W. W. Norton.

Rous, J. S. (2006). "The Body Dialogue Process." Maitland, FL: Unpublished manuscript.

Rubenfeld, I. (2001). *The listening hand: How to combine bodywork, intuition and psychotherapy to release emotions and heal the pain.* London, UK: Piatkus Books.

—— & Borysenko, J. (2001). *The listening hand: Self-healing through the Rubenfeld synergy method of talk and touch.* New York, NY: Bantam.

Rudnytsky, P. L. (2003). *Reading psychoanalysis: Freud, Rank, Ferenczi, Groddeck.* Ithaca, NY: Cornell University Press.

Russell, S. (2003). *The Tao of sexual massage: A step-by-step guide to exciting, enduring, loving pleasure.* New York, NY: Fireside.

Ruthven, M. (2007). *Fundamentalism.* London, UK: Oxford University Press.

Ryan, R. E. (2002). *Shamanism and the psychology of C. G. Jung: The great circle.* Falls Church, VA: Vega Publishing.

Sabetti, S. (2007). *Ki to psychology: A psychology and energy primer.* Sherman Oaks, CA: Life Energy Media.

Sabini, M. (ed., 2002). *The earth has a soul: The nature writings of C. G. Jung.* Berkeley, CA: North Atlantic Books.

Safran, J. D. (ed., 2003). *Psychoanalysis and Buddhism: An unfolding dialogue.* Boston, MA: Wisdom Publications.

Said, E. W. (1979). *Orientalism.* New York, NY: Vintage Books.

—— (1994). *Culture and imperialism.* New York, NY: Vintage Books.

Saillant, F. & Genest, S. (2007). *Medical anthropology. Regional perspectives and shared concerns.* Malden, MA: Blackwell.

Salguero, C. P. (2004). *Encyclopedia of Thai massage.* Findhorn, UK: Findhorn Press.

Saltonstall, E. (1988). *Kinetic Awareness: Discovering your bodymind.* New York, NY: Publishing Center Cultural Resources.

Samuels, A., Shorter, B. & Plaut, F. (2003). *A critical dictionary of Jungian analysis.* New York, NY: Brunner-Routledge.

Samuels, M. & Samuels, N. (1980). *Seeing with the mind's eye.* New York, NY: Random House.

Sandri, R. (1998). *Penser avec les bébés: Parcours, réflexions à partir de l'observation du bébé selon Esther Bick.* Paris, France: Erès.

Saraswati, S. S. (2001). *Kundalini tantra.* Coimbatore, India: Yoga Publishing Trust.

Sartre, J-P. 1956). *Being and nothingness: An essay on phenomenological ontology* (trans., H. E. Barnes). New York, NY: Philosophical Library.

—— (1965). *The transcendence of the ego: An existentialist theory of consciousness.* New York, NY: Noonday/Farrar, Straus and Giroux.

Scheper-Hughes, N. & Wacquant, L. (eds, 2003). *Commodifying bodies.* London, UK: Sage Publications.

Schneider, K. J. (1999). *The paradoxical self: Toward an understanding of our contradictory nature* (2nd ed.). Atlantic Highlands, NJ: Humanity Books.

—— (ed., 2007). *Existential-integrative psychotherapy: Guideposts to the core of practice.* New York, NY: Routledge.

——, Pierson, J. F. & Bugenthal, J. F. T. (eds, 2002). *The handbook of humanistic psychology: Leading edges in theory, research, and practice.* London, UK: Sage.

Schore, A. N. (1999). *Affect regulation and the origin of the self: The neurobiology of emotional development.* Hillsdale, NJ: Lawrence Erlbaum.

—— (2003). *Affect dysregulation and disorders of the self/Affect regulation and the repair of the self* (2 vols). New York, NY: W. W. Norton.

Schwartz-Salant, N. (2007). *The black nightgown: The fusional complex and the unlived life.* Wilmington, IL: Chiron Publications.

Scott, S. (1993). *Body matters: Essays on the sociology of the body.* London, UK: Routledge.

Scotton, B. W., Chinen, A. B. & Battista, J. R. (eds, 1996). *Textbook of transpersonal psychiatry and psychology.* New York, NY: Basic Books.

Seligman, M. E. P., Linley, P. A. & Joseph, S. (eds, 2004). *Positive psychology in practice.* New York, NY: Wiley.

Selver, C. & Brooks. C. V. W. (2007). *Reclaiming vitality and presence: Sensory awareness as a practice for life.* Berkeley, CA: North Atlantic Books.

Semin, G. R. & Smith, E. R. (eds, 2008). *Embodied grounding: Social, cognitive, affective, and neuroscientific approaches.* New York, NY: Cambridge University Press.

Sève, L. & McGreal, J. (1980). *Man in Marxist theory and the psychology of personality.* Upper Saddle River: Prentice Hall.

Sevilla, J. C. (2006). *Ecopsychology as ultimate force psychology: A biosemiotic approach to nature estrangement and nature alienation.* Philadelphia, PA: Xlibris/Random House.

Sharaf, M. (1994). *Fury on earth: A biography of Wilhelm Reich.* Cambridge, MA: Da Capo Press.

Sheikh, A. A. & Sheikh, K. S. (1996). *Healing east and west: Ancient wisdom and modern psychology.* New York, NY: Wiley.

Sheldon, W. H. (1940). *The varieties of human physique: An introduction to constitutional psychology.* New York, NY: Harper & Brothers.

Shepher, J. (1983). *Incest: A biosocial view.* Burlington, MA: Academic Press.

Shilling, C. (2003). *The body and social theory* (2nd ed.). London, UK: Sage Publications.

Shannahoff-Khalsa, D. (2007). *Kundalini meditation: Techniques specific for psychiatric disorders, couples therapy and personal growth.* New York, NY: W. W. Norton.

Shorter, E. (1998). *A history of psychiatry: From the era of the asylum to the age of Prozac* (2nd ed.). New York, NY: Wiley.

Siegel, D. J. (1999). *The developing mind: How relationships and the brain interact to shape who we are.* New York, NY: Guilford Press.

—— (2007). *The mindful brain: Reflection and attunement in the cultivation of well-being.* New York, NY: W. W. Norton.

Siegel, E. (1984). *Dance movement therapy: Mirror of ourselves: The psychoanalytic approach.* New York: Human Science Press.

Skinner, B. F. (1978). *Reflections on behaviorism and society.* Upper Saddle River, NJ: Prentice Hall.

Sloan, T. (1996). *Damaged life: The crisis of the modern psyche.* New York, NY: Routledge.

—— (ed., 2000). *Critical psychology: Voices for change*. New York, NY: St. Martin's Press.

Smith, E. W. L. (2003). *The body in psychotherapy*. Jefferson, NC: McFarland & Co.

Sontag, S. (2001). *Illness as metaphor (and AIDS and its metaphors)*. New York, NY: Picador.

Spiegelberg, H. (1972). *Phenomenology in psychology and psychiatry: A historical introduction*. Evanston, IL: Northwestern University Press.

Spivak, G. C. (1999). *A critique of postcolonial reason: Toward a history of the vanishing present*. Cambridge, MA: Harvard University Press.

Staddon, J. (2001). *The new behaviorism: Mind, mechanism and society*. Philadelphia, PA: Psychology Press.

Standring, S. (ed., 2008). *Gray's anatomy: The anatomical basis of clinical practice* 40th ed.). Edinburgh, UK: Churchill Livingstone.

Stanton-Jones, K. (1992). *An introduction to dance movement therapy in psychiatry*. London, UK: Tavistock/Routledge.

Stapp, H. P. (2007). *Mindful universe: Quantum mechanics and the participating observer*. New York, NY: Springer.

—— (2009). *Mind, matter and quantum mechanics* (3rd ed.). New York, NY: Springer.

Stattman, J. (ed., 1989). *Unitive bodypsychotherapy: Collected papers*. Butzbach, Germany: Afra Verlag.

Stein, D. (1995). *Essential reiki: A complete guide to an ancient healing art*. New York, NY: Crossing Press.

Stein, R. (1984). *Incest and human love: The betrayal of the soul in psychotherapy*. Dallas, TX: Spring Publications.

Steiner, R. & Usher, B. (2007). *Eurythmy: An introductory reader*. Bucharest, Romania: Sophia Books.

Still, A. T. (1992). *Osteopathy: Research and practice*. Seattle, WA: Eastland Press.

—— (2009). *Philosophy of osteopathy*. Brooklyn, NY: AMS Press Inc.

Stone, H. & Stone, S. (1998). *Embracing ourselves: The voice dialogue manual* (new edition). Novato, CA: New World Library/Nataraj.

Stubbs, K. R. (ed., 1994). *Women of the light: The new sacred prostitute*. Grawn, MI: Access Publishers Network.

—— (1999). *Erotic massage*. New York, NY: Tarcher.

Stutley, M. (2002). *Shamanism: An introduction*. New York, NY: Routledge.

Suzuki, D. T. (1996). *Zen Buddhism: Selected writings of D. T. Suzuki* (ed., W. Barrett). New York, NY: Three Leaves Press.

——, Fromm, E. & DeMartino, R. (eds, 1960). *Zen Buddhism and psychoanalysis*. New York, NY: Harper & Row.

Sweigard, L. E. (1988). *Human movement potential: Its ideokinetic facilitation*. Lanham, MD: University Press of America.

Szasz, T. S. (1984). *The myth of mental illness: Foundations of a theory of personal conduct* (revised edition). New York, NY: Harper Paperbacks.

—— (1988). *The myth of psychotherapy: Mental healing as religion, rhetoric, and repression*. Syracuse, NY: Syracuse University Press.

—— (1989). *Law, liberty, and psychiatry: An inquiry into the social uses of mental health practices*. Syracuse, NY: Syracuse University Press.

—— (2007a). *Coercion as cure: A critical history of psychiatry.* New York, NY: Transaction Books.

Tagore, R. (1962). *Collected poems and plays of Rabindranath Tagore.* London, UK: Macmillan.

Tansley, D. (1984). *Chakras: Rays and radionics.* Saffron Walden, UK: C. W. Daniel.

—— (1985). *The raiment of light: A study of the human aura.* London, UK: Routledge.

Targ, R. (2004). *Limitless mind: A guide to remote viewing and transformation of consciousness.* Novato, CA: New World Library.

Tart, C. T. (1983). *States of consciousness.* El Cerrito, CA: Psychological Processes.

—— (ed., 1992). *Transpersonal psychologies: Perspectives on the mind for seven great spiritual traditions.* New York, NY: HarperCollins.

Taylor, M. C. (1987). *Altarity.* Chicago, IL: University of Chicago Press.

—— (1993). *Nots.* Chicago, IL: University of Chicago Press.

Teeguarden, I. M. (2002). *A complete guide to acupressure: Jin Shin Do.* Tokyo, Japan: Japan Publications.

Teo, T. (2005). *The critique of psychology: From Kant to postcolonial theory.* New York, NY: Springer.

Thomas (Judas) (1992). *The gospel of Thomas: The hidden sayings of Jesus* (trans. & ed., M. Meyer). San Francisco, CA: HarperSanFrancisco.

Toadvine, T. (2009). *Merleau-Ponty's philosophy of nature.* Evanston, IL: Northwestern University Press.

Todd, J. T. & Morris, E. K. (1994). *Modern perspectives on John B. Watson and classical behaviorism.* Westport, CT: Greenwood Press.

Todd, M. (1980). *The thinking body.* Princeton, NJ: Princeton Book Company.

Todes, S. (2001). *Body and world.* Cambridge, MA: M.I.T. Press.

Tolman, C. W. (1994). *Psychology, society, and subjectivity: An introduction to German critical psychology.* London, UK: Routledge.

Totten, N. (1998). *The water in the glass: Body and mind in psychoanalysis.* London, UK: Rebus.

—— (2003). *Body psychotherapy: An introduction.* Maidenhead, UK: Open University Press.

—— (ed., 2005). *New dimensions in body psychotherapy.* Maidenhead, UK: Open University Press.

Trungpa, C. (2001). *Crazy wisdom.* Boston, MA: Shambhala.

—— (2004). *Meditation in action.* Boston, MA: Shambhala.

—— (2007). *Shambhala: The sacred path of the warrior.* Boston, MA: Shambhala.

Tsering, G. T. & Rinpoche, Z. T. (2006). *Buddhist psychology: The foundation of Buddhist thought.* Boston, MA: Wisdom Publications.

Turner, B. S. (2000). *The Blackwell companion to social theory* (2nd ed.). Oxford, UK: Wiley-Blackwell.

—— (2008). *The body and society: Explorations in social theory.* London, UK: Sage Publications.

Turner, J. H. & Maryanski, A. (2005). *Incest: Origins of the taboo.* Boulder, CO: Paradigm Publishers.

Tutu, D. (1999). *No future without forgiveness.* New York, NY: Random House.

Ueshiba, M. (1999). *The essence of aikido: Spiritual teachings of Morihei Ueshiba.* Tokyo, Japan: Kodansha International.

—— & Ueshiba, M. (2008). *The secret teachings of aikido*. Tokyo, Japan: Kodansha International.

Unno, M. (2006). *Buddhism and psychotherapy across cultures: Essays on theories and practices*. Boston, MA: Wisdom Publications.

Unschuld, P. U. (1988). *Medicine in China: A history of ideas*. Berkeley, CA: University of California Press.

Usui, M. & Petter, F. A. (1999). *The original reiki handbook*. Twin Lakes, WI: Lotus Press.

Valle, R. S. & Eckartsberg, R. V. (eds, 1981). *The metaphors of consciousness*. New York, NY: Plenum Press.

Van der Kolk, B. A. (1987). *Psychological trauma*. Arlington, VA: American Psychiatric Press.

—— (1994). *The body keeps the score: Memory and the evolving psychobiology of posttraumatic stress*. Cambridge, MA: Harvard Medical.

——, McFarlane, A. C. & Weisaeth, L. (eds, 1996). *Traumatic stress: The effects of overwhelming experience on mind, body, and society*. New York, NY: Guilford Press.

Van Kaam, A. L. (1960). *The third-force in European psychology: Its expression in a theory of psychotherapy*. New York, NY: Psychosynthesis Research Foundation.

Varela, F. J., Thompson, E. T. & Rosch, E. (1992). *The embodied mind: Cognitive science and human experience*. Cambridge, MA: M.I.T. Press.

Vigarello, G. (2005). *Histoire du corps*. Paris, France: Seuil.

Villoldo, A. (2000). *Shaman, healer, sage: How to heal yourself and others with the energy medicine of the Americas*. Van Nuys, CA: Harmony.

—— (2005). *Mending the past and healing the future with soul retrieval*. Carlsbad, CA: Hay House.

—— (2006). *Courageous dreaming: How shamans dream the world into being*. Carlsbad, CA: Hay House.

—— & Jendresen, E. (1994). *Dance of the four winds: Secrets of the Inca medicine wheel*. Rochester, VT: Destiny Books.

—— & Krippner, S. (1987). *Healing states: A journey into the world of spiritual healing and shamanism*. New York, NY: Fireside.

Vitebsky, P. (2001). *Shamanism*. Norman, OK: University of Oklahoma Press.

Vivekananda, S. (1947). *Complete works of Swami Vivekananda* (8 vols). Hollywood, CA: Vedanta Press.

—— (2007). *The Yoga-Sûtras of Patanjali: The essential yoga texts for spiritual enlightenment*. London, UK: Watkins Publishing.

Vries, H. de (2005). *Minimal theologies: Critiques of secular reason in Adorno and Levinas* (trans. G. Hale). Baltimore, MD: Johns Hopkins University Press.

Wallace, B. A. (2007). *Contemplative science: Where Buddhism and neuroscience converge*. New York, NY: Columbia University Press.

Walsh, R. (2007). *The world of shamanism: New views of an ancient tradition*. Woodbury, MN: Llewellyn Publications.

—— & Grob, C. (eds, 2005). *Higher wisdom: Eminent elders explore the continuing impact of psychedelics*. Albany, NY: State University of New York Press.

—— & Vaughan, F. (eds, 1993). *Paths beyond ego: The transpersonal vision*. New York, NY: Tarcher Putnam.

Walter, M. & Fridman, E. (2004). *Shamanism: An encyclopedia of world beliefs, practices, and culture* (2 vols). Santa Barbara, CA: ABC-CLIO.

Wang, Y. (2001). *Buddhism and deconstruction: Toward a comparative semiotics.* London, UK: Routledge Curzon.

Ward, R. C., Hruby, R. J., Jerome, J. A., Jones, J. M. & Kappler, R. E. (2002). *Foundations for osteopathic medicine* (2nd ed.). Philadelphia, PA: Lippincott, Williams & Wilkins.

Washburn, M. (1994). *Transpersonal psychology in psychoanalytic perspective.* Albany, NY: State University of New York Press.

—— (1995). *The ego and the dynamic ground: A transpersonal theory of human development* (2nd ed.). Albany, NY: State University of New York Press.

Watkins, M. (1986). *Invisible guests: The development of imaginal dialogues.* Hillsdale, NJ: Analytic Press.

—— & Shulman, H. (2008). *Toward psychologies of liberation.* New York, NY: Palgrave Macmillan.

Watson, G. (1998). *The resonance of emptiness: A Buddhist inspiration for contemporary psychotherapy.* New York, NY: Columbia University Press.

Watson, J. B. (1930). *Behaviorism* (revised edition). Chicago, IL: University of Chicago.

Watts, A. W. (1961). *Psychotherapy east and west.* New York, NY: Pantheon Books.

—— (1991). *Nature, man and woman.* New York, NY: Vintage.

—— (1999). *The way of Zen.* New York, NY: Vintage.

Waugh, A. & Grant, A. (2001). *Anatomy and physiology in health and illness.* Edinburgh, UK: Churchill Livingstone.

Webb, H. S. (2004). *Traveling between the worlds: Conversations with contemporary shamans.* Charlottesville, VA: Hampton Roads Publishing.

—— (2008). *Exploring shamanism.* Franklin Lakes, NJ: Career Press.

Weiser, A. (1996). *The power of focusing: A practical guide to emotional self-healing.* Oakland, CA: New Harbinger.

Weiss, G. & Haber, H. F. (eds, 1999). *Perspectives on embodiment: The intersections of nature and culture.* New York, NY: Routledge.

Weitz, R. (2002). *The politics of women's bodies: Sexuality, appearance, and behavior.* London, UK: Oxford University Press.

Welwood, J. (1979). *The meeting of the ways: Explorations in east/west psychology.* New York, NY: Schocken Books.

—— (2002). *Toward a psychology of awakening: Buddhism, psychotherapy, and the path of personal and spiritual transformation.* Boston, MA: Shambhala Press.

Wesselman, H. (2003). *Journey to the sacred garden: A guide to traveling in the spiritual realms.* Carlsbad, CA: Hay House.

—— (2004). *Spirit medicine: Healing in the sacred realms.* Carlsbad, CA: Hay House.

Wiener, D. J. (ed., 2001). *Beyond talk therapy: Using movement and expressive techniques in clinical practice.* Washington, DC: American Psychological Association.

Wilber, K. (ed., 1985). *Quantum questions: Mystical writings of the world's great physicists.* Boston, MA: Shambhala.

—— (1990). *Eye to eye: The quest for a new paradigm.* Boston, MA: Shambhala.

—— (1998). *The marriage of sense and soul: Integrating science and religion.* New York, NY: Random House.

—— (2000). *Integral psychology: Consciousness, spirit, psychology, therapy.* Boston, MA: Shambhala Press.

Wilce, J. M. (ed., 2003). *Social and cultural lives of immune systems.* New York, NY: Routledge.

Wilden, A. (1968). *The language of the self.* Baltimore, MD: Johns Hopkins University Press.

—— (1972). *System and structure: Essays in communication and exchange.* London, UK: Tavistock.

Wilson, R. A. (2004). *Boundaries of the mind: The individual in the fragile sciences.* London, UK: Cambridge University Press.

Winston, A. (ed., 2004). *Defining difference: Race and racism in the history of Psychology.* Washington, DC: American Psychological Association.

Winter, D. D. (1996). *Ecological psychology: Healing the split between planet and self.* New York, NY: HarperCollins.

Wood, D. (2001). *The deconstruction of time.* Evantson, IL: Northwestern University Press.

—— (2007). *Time after time.* Bloomington, IN: Indiana University Press.

Woodhull, V. C. (2005). *And the truth shall make you free.* London, UK: Dodo Press.

Wujastyk, D. (ed., 2002). *The roots of Āyurveda: Selections from Sanskrit medical writings.* New York, NY: Penguin.

Yao, Z. (2005). *The Buddhist theory of self-cognition.* London, UK: Routledge.

Young-Bruehl, E. (1998). *The anatomy of prejudices.* Cambridge, MA: Harvard University Press.

Zaner, R. (1964). *The problem of embodiment: Some contributions to a phenomenology of the body.* The Hague, Netherlands: Martinus Nijhoff.

Zhang, Yanhua (2007). *Transforming emotions with Chinese medicine: An ethnographic account from contemporary China.* Albany, NY: State University of New York Press.

Zhang, Yuanxia (2006). *Zen and psychotherapy.* Victoria, Canada: Trafford Publishing.

Zimbardo, P. G. (2008). *The lucifer effect: Understanding how good people turn evil.* New York, NY: Random House.

Znamenski, A. A. (2007). *The beauty of the primitive: Shamanism and western imagination.* London, UK: Oxford University Press.

Zohar, D. (1990). *The quantum self: Human nature and consciousness defined by the new physics.* New York, NY: William Morrow.

About the Author

Barnaby B. Barratt, PhD, DHS, earned his first doctorate in Psychology and Social Relations from Harvard University and his second from the Institute for Advanced Study of Human Sexuality. He completed clinical training at the University of Michigan's Neuropsychiatric Institute as well as at the Michigan Psychoanalytic Institute, and for many years he was Professor of Family Medicine, Psychiatry and Behavioral Neurosciences at Wayne State University in Detroit, where he directed programs in human sexuality, mental health and community medicine. An elected Fellow of the American Psychological Association and of the American Academy for Psychoanalysis in Psychology, as well as a diplomate of the American Board of Professional Psychology, Dr. Barratt is an active member of the International Psychoanalytic Association and the United States Association for Body Psychotherapy. He has trained intensively in a variety of bodymind modalities, including Ashtanga Yoga, Thai massage, and meditation in the tantric tradition. As a psychoanalyst, sexuality educator and sex therapist, Dr. Barratt practices somatic and psychodynamic therapy. He currently maintains a private clinand educational practice, and he offers seminars and workshops throughout the United States and internationally. He can be contacted *BBBarratt@Earthlink.net.*

Index

CPSIA information can be obtained at www.ICGtesting.com
Printed in the USA
LVOW10s1700240116

472074LV00008B/97/P